About Island Press

Since 1984, the nonprofit Island Press has been stimulating, shaping, and communicating the ideas that are essential for solving environmental problems worldwide. With more than 800 titles in print and some 40 new releases each year, we are the nation's leading publisher on environmental issues. We identify innovative thinkers and emerging trends in the environmental field. We work with world-renowned experts and authors to develop cross-disciplinary solutions to environmental challenges.

Island Press designs and implements coordinated book publication campaigns in order to communicate our critical messages in print, in person, and online using the latest technologies, programs, and the media. Our goal: to reach targeted audiences—scientists, policymakers, environmental advocates, the media, and concerned citizens—who can and will take action to protect the plants and animals that enrich our world, the ecosystems we need to survive, the water we drink, and the air we breathe.

Island Press gratefully acknowledges the support of its work by the Agua Fund, Inc., Annenberg Foundation, The Christensen Fund, The Nathan Cummings Foundation, The Geraldine R. Dodge Foundation, Doris Duke Charitable Foundation, The Educational Foundation of America, Betsy and Jesse Fink Foundation, The William and Flora Hewlett Foundation, The Kendeda Fund, The Andrew W. Mellon Foundation, The Curtis and Edith Munson Foundation, Oak Foundation, The Overbrook Foundation, the David and Lucile Packard Foundation, The Summit Fund of Washington, Trust for Architectural Easements, Wallace Global Fund, The Winslow Foundation, and other generous donors.

The opinions expressed in this book are those of the author(s) and do not necessarily reflect the views of our donors.

GREENING OUR
BUILT WORLD

GREENING OUR BUILT WORLD

COSTS, BENEFITS, AND STRATEGIES

GREG KATS,
Principal Author

JON BRAMAN & MICHAEL JAMES

ISLANDPRESS
WASHINGTON | COVELO | LONDON

Library of Congress Cataloging-in-Publication Data

Kats, Gregory.
 Greening our built environment : costs, benefits, and strategies / Greg Kats, Principal
author ; Jon Braman and Michael James.
     p. cm.
 Includes bibliographical references and index.
 ISBN-13: 978-1-59726-667-3 (cloth : alk. paper)
 ISBN-10: 1-59726-667-1 (cloth : alk. paper)
 ISBN-13: 978-1-59726-668-0 (pbk. : alk. paper)
 ISBN-10: 1-59726-668-X (pbk. : alk. paper) 1. Sustainable architecture--Economic
aspects. 2. Sustainable design--Economic aspects. 3. Green technology--Economic as-
pects. I. Braman, Jon. II. James, Michael, 1939- III. Title.
 NA2542.36.K38 2009
 720'.47--dc22

                                                                          2009026848

Printed on recycled, acid-free paper ⊛

Design by Joan Wolbier

Manufactured in the United States of America
10 9 8 7 6 5 4 3 2

In loving memory of
Tuckerman, Evelina, and Ivan Kats.
And with thanks to my wife, Maia,
for her support, love, and wisdom.

CONTENTS

XI CONTRIBUTORS

XII SPONSORING ORGANIZATIONS

XIII ACKNOWLEDGMENTS

XV INTRODUCTION

1 PART I: COSTS AND BENEFITS OF GREEN BUILDING

3 1.1 Methodology

8 1.2 The Cost of Building Green

14 1.3 Energy-Use Reductions

26 1.4 Advanced Energy-Use Reductions

33 1.5 Water-Related Savings

40 1.6 Green Affordable Housing: Enterprise's Green Communities Initiative

46 1.7 Health and Productivity Benefits of Green Buildings

59 1.8 Green Health Care: Assessing Costs and Benefits

66 1.9 Employment Benefits of Green Buildings

73 1.10 Property Value Impacts of Building Green

84 1.11 Net Financial Impacts of Green Buildings for Owners and Occupants

89 PART II: COSTS AND BENEFITS OF GREEN COMMUNITY DESIGN

93 2.1 What Is a Green Community?

98 2.2 Setting the Stage for Sustainable Urbanism

103 2.3 Financial Impacts of Green Community Design

110 2.4 Transportation and Health Impacts of Green Community Design

124 2.5 Property Value and Market Impacts

127 2.6 The Market Rediscovers Walkable Urbanism

131 2.7 Social Impacts of Green Communities

134 2.8 Cost Savings in Ecologically Designed Conservation Developments

141 2.9 International Green Building

147 2.10 Financial Impact of Green Communities

149 PART III: COMMUNITIES OF FAITH BUILDING GREEN

151 3.1 Faith Groups in the Green Vanguard

154 3.2 Methodology and Findings

160 3.3 Motivation

165 3.4 Impact of Green Buildings in Faith Communities

168 3.5 Financial Stewardship

169 3.6 Conclusion

171   PART IV: GREEN DESIGN, CLIMATE CHANGE,
         AND THE ECONOMY: POTENTIAL
         IMPACTS IN THE UNITED STATES

   175   4.1 Energy Consumption

   181   4.2 Renewable Energy

   183   4.3 Carbon Dioxide Emissions

   187   4.4 Financial Impact

189   CONCLUSION

PERSPECTIVES

    32   PERSPECTIVE: BANK OF AMERICA
         TOWER AT ONE BRYANT PARK

    35   PERSPECTIVE: WATER-SAVING
         STRATEGIES: OREGON HEALTH
         SCIENCES UNIVERSITY CENTER FOR
         HEALTH AND HEALING

    53   PERSPECTIVE: BIRTH OF THE GREEN
         BRANCH BANK

    72   PERSPECTIVE: GREEN BUILDING AS
         CORPORATE SOCIAL RESPONSIBILITY

    79   PERSPECTIVE: INVESTING IN
         BROWNFIELDS

    83   PERSPECTIVE: MEASURING CONSUMER
         DEMAND FOR GREEN HOMES

   121   PERSPECTIVE: MONITORING
         PERFORMANCE AT THE LEED
         PLATINUM CENTER FOR
         NEIGHBORHOOD TECHNOLOGY

   142   PERSPECTIVE: GREEN BUILDINGS IN
         CHINA

   146   PERSPECTIVE: A GREENER ECONOMIC
         RECOVERY

APPENDIXES

   191   APPENDIX A: DATA-COLLECTION
         METHODOLOGY

   195   APPENDIX B: SOURCE LIST

   199   APPENDIX C: GREEN BUILDING
         DATA SET

   212   APPENDIX D: COMPARISON OF DATA
         SET TO LEED–NEW CONSTRUCTION
         BUILDINGS

   214   APPENDIX E: BASELINES USED IN
         COST AND BENEFITS ESTIMATES

   215   APPENDIX F: ISSUES IN RESEARCHING
         THE COST OF GREEN BUILDING

   217   APPENDIX G: COST OF ENERGY-
         EFFICIENCY AND RENEWABLE-ENERGY
         MEASURES

   218   APPENDIX H: ENERGY-USE
         BASELINES AND STANDARDS

   220   APPENDIX I: VERIFYING THE ENERGY
         PERFORMANCE OF LEED BUILDINGS

   222   APPENDIX J: ASSUMPTIONS USED FOR
         CALCULATIONS OF WATER SAVINGS

   225   APPENDIX K: GREEN BUILDING
         SURVEY INSTRUMENT

   226   APPENDIX L: GLOBAL ASSUMPTIONS
         FOR PART IV

229   ABOUT THE AUTHORS

233   NOTES

251   INDEX

CONTRIBUTING AUTHORS

Steven I. Apfelbaum, Dana Bourland, Tom Darden, Jill C. Enz, Douglas Farr, Robert F. Fox Jr., Lawrence Frank, Robin Guenther, Adele Houghton, Sarah Kavage, John (Skip) Laitner, Joe Lehman, Christopher B. Leinberger, Gary Jay Saulson, Craig Q. Tuttle, Stockton Williams, Gail Vittori, and Sally Wilson

ADDITIONAL CONTRIBUTORS

Alex Buell, Father Drew Christiansen, Rachel Dewane, Kathryn Eggers, Breeze Glazer, Shyam Kannan, Paul King, Melanie Simmons, Rachel Scheu, and Rob Watson

SPONSORING ORGANIZATIONS

American Council on Renewable Energy (www.acore.org)

American Institute of Architects (www.aia.org)

American Public Health Association (www.apha.org)

BOMA International (www.boma.org)

Enterprise Community Partners (www.enterprisecommunity.org)

Federation of American Scientists (www.fas.org)

National Association of Realtors (www.realtor.org)

National Association of State Energy Officials (www.naseo.org)

Real Estate Roundtable (www.rer.org)

U.S. Green Building Council (www.usgbc.org)

World Green Building Council (www.worldgbc.org)

ACKNOWLEDGMENTS

The authors gratefully acknowledge the generosity of Adam J. Lewis in funding the research for this book.

We are deeply grateful to Good Energies for its sustained support, and would like to particularly thank Richard Kauffman and Marcel Brenninkmeijer for their vision, leadership, and support.

Thank you, Teddy Goldsmith, Denis Hayes, Amory Lovins, David Orr and Art Rosenfeld, for a quarter century of friendship, inspiration and collaboration.

The authors wish to thank the following reviewers: Andrew Aurbach, Alex Buell (Good Energies), Katherine Hamilton (Gridwise Alliance), Emely Lora (Good Energies), Alex Kats-Rubin, Kathleen Mahoney (Humanitas Foundation), Michelle Riley (Good Energies) and Joe Romm (Center for American Progress).

Many individuals and organizations lent their insight and time to the research and development of this book. The authors gratefully acknowledge the contributions of the following: Al Nichols, Al Nichols Engineering; Julian Dautremont-Smith, Association for the Advancement of Sustainability in Higher Education; John (Skip) Laitner and Steve Nadel, American Council for an Energy-Efficient Economy; Michael Eckhart, American Council on Renewable Energy; Frank Loy; Marcus Chang, Applied Ecological Services; Peter Mayer, Aquacraft; Mary McLeod, Austin Green Building Program; Jonathan Cahn, Baker & McKenzie; Bob Berkebile, BNIM; Peter Fox-Penner, Brattle Group; Nadav Malin, Michael Wentz, and Alex Wilson, Building Green; Vivian Loftness, Carnegie Mellon University; Gail Braeger, Center for the Built Environment; Jan Hamrin and Jennifer Martin, Center for Resource Solutions/Green E; Howard Frumkin, Centers for Disease Control; Tom Darden and Chris Wedding, Cherokee Development Partners; John Norquist, Congress for New Urbanism; Glenn Prickett, Conservation International; Jay Spivey, CoStar; Andrew Russell, CPB; Lisa Matthiessen, Davis Langdon; Elizabeth Plater-Zyberk and Michael Watkins, Duany Plater-Zyberk, Inc.; John Manning, Earth Sensitive Solutions; Rob Watson, EcoTech International; Dana Bourland, Enterprise Community Partners; Douglas Farr, Farr Associates; Steve Bushnell and David Cohen, Fireman's Fund Insurance Company; Dennis Whittle, Global Giving; Michael Ware, George Coelho, Patrick Flynn, and Pat Sapinsley, Good Energies; Jan Hamrin and Jennifer Martin, Center for Resource Solutions/Green E; Judith Heerwagen, Heerwagen & Associates; Rashad Kaldany, IFC; Dr. Nicolas Kats; John Gattuso,

Liberty Property Trust; John Boecker, L. Robert Kimball & Associates: Architects and Engineers; Jeff Genzer, National Association of State Energy Officials; Mark Frankel and Cathy Turner, New Buildings Institute; Neil MacFarquhar, Frances Beineke and Ashok Gupta, NRDC; David Orr, Oberlin College; Theddi Chappell, Pacific Security Capital; Tom Paladino and Brad Pease, Paladino and Company; Katie McGinty, Peregrine Technology Partners; Jonathan Spalter, Public Insight, LP; Peiffer Brandt, Raftelis Financial Consultants; Greg Franta and Amory Lovins, Rocky Mountain Institute; Hank Habicht, Sail Ventures; Marcus Sheffer, Seven Group; Michael Saxenian and Bruce Stewart, Sidwell Friends School; Steve Winter, Steve Winter Associates; Michael Mehaffey, Structura Naturalis, Inc.; Bill Browning, Terrapin; Laurence Aurbach, the *Town Paper,* Pedshed.net; Peter Banwell, U.S. Environmental Protection Agency; Tom Dietsche, Rick Fedrizzi, Doug Gatlin, Jennifer Henry, Tom Hicks, Scott Horst, Michelle Moore, Brendan Owens, and Chris Smith, U.S. Green Building Council; Sam Baldwin, Dru Crawley, Richard Duke, Mark Ginsberg, Henry Kelly, and David Sandalow, U.S. Department of Energy; Brenna Walraven, USAA Real Estate Company; Neil Chambers, Vanderbilt College; Amy Vickers, Vickers and Associates; Steve Mufson, the *Washington Post;* Joshua Horwitz, Waterford Life Sciences/Living Planet Books; Melissa Ferrato and Kevin Hyde, World Green Building Council; and thank you, Tim Foote, Tom Lovejoy, and Jonathan Spalter.

Finally, we would like to extend sincere thanks to Island Press and its staff for the invaluable editorial, production and marketing guidance and support throughout this process. In particular, Sandy Chizinsky's thoughtful copyediting, Marcia Rackstraw's graphic design expertise, and Sharis Simonian's skill as a production editor all helped to make our vision for this book a reality.

ABOUT GOOD ENERGIES

Greg Kats is Senior Director and Director of Climate Change Policy at Good Energies, a leading global investor in renewable energy and energy efficiency industries with several billion dollars under management. The firm invests in solar, turbine-based renewables, green building technologies and other emerging areas in clean energy. Founded in 2001, Good Energies manages the renewable energy portfolio of COFRA, a family owned and managed group of companies. Guided by the "3-P" principle of People-Planet-Profit, Good Energies makes long-term investments in companies with outstanding growth potential. The firm's mission is to accelerate the global transition to a low-carbon economy.

INTRODUCTION

We shape our buildings,
and afterwards our buildings shape us.

—WINSTON CHURCHILL

Construction is how we shape and reshape our physical world. Buildings—our homes, schools, offices, cities, and towns—define where and how we live and work, and how we use resources, including almost half the energy we consume. Buildings are also a major part of the legacy we leave our children. Nevertheless, buildings are typically designed and constructed to meet cost objectives, with little thought to how they relate to each other or how they shape our lives and livelihoods. As a result, buildings separate as much as they link us, locking us into patterns of consumption that are neither healthy nor environmentally sustainable. Green design offers a new direction.

Green buildings—designed to use fewer resources and to support the health of their inhabitants—are commonly viewed as more expensive to build than conventional buildings. For example, a 2007 opinion survey by the World Business Council for Sustainable Development found that, on average, green buildings were thought to cost 17 percent more than conventional buildings. However, we found this widespread perception—that greening costs a lot more than conventional design—to be wrong. In fact, the 170 green buildings analyzed for this book cost, on average, only 2% more than conventional buildings; moreover, green buildings provide a wide range of benefits—both direct and indirect—that typically make them a very good investment.

The global recession begun in 2008 was triggered by collapsing home values and marked by a deep slowdown in construction. Green design continued to grow, though more slowly. Although only 1% of existing buildings in the United States are green, substantial anecdotal information suggests that green buildings and homes command higher rents and sales prices. In addition, a growing number of public institutions require or give preference to green design, and demand for green retrofits is increasing rapidly.

GOALS AND APPROACH

We wrote this book to explore the broader potential for green design, and to answer the fundamental question of whether the benefits of green design outweigh the costs. The answer will largely determine whether green design can make the transition from environmentally motivated niche to cost-conscious mainstream. And, critically, if green design is broadly cost-effective, how large an impact could greening have on shifting to a clean energy economy and slowing global warming?

Part I provides a framework for estimating the magnitude of the costs and benefits of individual green buildings. We assembled detailed data on 170 recent green buildings—the largest and most extensive analysis to date—to quantify the diverse benefits of building green, including energy and water savings, health and productivity improvements, job creation, and property value increases. We also provide an in-depth discussion of the costs and benefits of 18 buildings projected to use at least 50% less energy than similar conventional buildings.

Communities—ranging from houses of worship to universities to neighborhoods—are adopting green design not just for financial reasons but to strengthen and reaffirm their values and to support the health and well-being of the people they serve. These larger social and environmental impacts, though difficult to measure, increasingly influence design and development choices. Parts II and III are about greening two different types of communities: neighborhood-scale residential developments and religious communities. Part II reviews and evaluates green community development, including the costs of site development and storm-water infrastructure, and impacts on energy and water use, health, transportation, and property value. As with individual green buildings, there is a perception that green communities cost more. But an analysis developed for this book, of ten recent conservation developments comprising more than 1,500 homes, shows an average of $12,000 in first-cost savings per home site, largely because of reduced infrastructure costs, including costs for water treatment.

Part III presents the results of a more qualitative survey of 17 faith-based institutions that have built green buildings. For a growing number of religious institutions, building green has become not just a cost-effective investment but, more importantly, a way to embody and demonstrate a religious and moral commitment to care for the earth and for life. The process of learning about and undertaking greening, in turn, commonly reinvigorates the religious community.

Part IV develops two national building scenarios through 2050: a business-as-usual scenario and a green scenario. This maps the potential for a national shift to green design as a means of cutting energy dependence, achieving national financial savings, and slowing global warming. The business-as-usual scenario is based on Energy Information Agency projections, but assumes a somewhat more rapid growth in energy efficiency, green buildings, and renewable energy. In this scenario, new green construction increases fivefold but maxes out as a large niche market equal to 25% of construction. In the green scenario, energy efficiency and green construction and retrofitting spread more rapidly, and become the norm, driving deep reductions in energy use and carbon dioxide (CO_2)

emissions from the building sector. The findings in Part IV indicate that a rapid and sustained national transition to green design is both cost-effective and feasible.

Throughout the book, perspective pieces from leading practitioners in the field of green design—architects, developers, researchers, property owners—share their own experiences in greening institutions—ranging from banks to affordable housing and from religious institutions to residential developments. A common lesson emerges: green design is cost-effective, *and* it creates important additional benefits—such as strengthening a community or reaffirming a church's purpose—that may be unquantifiable but are no less important.

BUILDING GREEN: COSTS, BENEFITS, AND POTENTIAL

The cost of green building is minimal—and makes for a very good investment. From energy savings alone, the average payback time for a green building is six years. Additional benefits include reduced water and infrastructure costs, and health and productivity gains; these benefits more than double the financial gains for green building owners and occupants. Over 20 years, the financial payback commonly exceeds the additional cost of greening by a factor of between four and six. And broader benefits, such as reductions in greenhouse gases (GHGs) and pollution, have large positive impacts on surrounding communities and on the planet.

If energy prices rise at 5% per year (which is below the rate at which energy prices grew from 2004 to 2008), then, over 20 years, energy savings are twice the cost of greening. But if energy costs rise faster—say, at 8% per year—energy savings would be over three times the average cost of greening. The volatility of energy prices and the long-term trend of rising demand for finite and depleting fossil fuels make greening and energy efficiency cost-effective risk-reduction strategies.

Green building also creates more jobs than conventional construction. Energy efficiency, renewable energy, and waste diversion (e.g., separation and recycling) are all common features of green buildings—and all provide significantly more employment than conventional design, while greater efficiency and the use of renewable energy cut reliance on imported fossil fuels. Energy-efficient construction, for example, requires more time insulating and caulking walls, roofs, and basements. But the higher cost of this additional work is offset by reduced energy waste and by long-term reductions in the purchase of energy, some of which is imported. Moreover, many of the jobs created by a shift to green design require specialized skills, and lead to good permanent local jobs.

Venture-capital investment in clean energy increased tenfold between 2003 and 2008, accelerating the development and deployment of technologies that allow buildings to be far more energy- and water-efficient, and that make it possible to cost-effectively generate energy on site. The increasingly rapid development of highly efficient lighting, windows, motors, and controls makes deep cuts in energy use feasible. The rapid growth of the green building industry has accelerated this trend. Of the 170 buildings analyzed in this book, 18 projected energy-efficiency cuts of 50% or more; most of these buildings featured on-site renewable power generation—a pattern that points toward a future of zero-energy buildings.

The effect of buildings on the environment, including human health, is substantially determined by where buildings are located in relation to each other, and in relation to open space, public transport, and other amenities. Clustering development leaves much more open land and increases residents' access to fields, trails, and woods—all of which increase property value. At its best, green community design fosters non-automobile forms of transportation (walking, biking, and transit) and helps to create diverse, socially vibrant neighborhoods with a rich mix of social and commercial activities. Higher densities and ready access to amenities give residents more destinations to walk to, and can substantially cut both driving and pollution—and their associated personal and societal costs.

In contrast to conventional sprawl, green design supports stable communities instead of inhibiting them. Alexis de Tocqueville observed 175 years ago, in *Democracy in America*, that one of America's enduring strengths is the tendency to form resilient voluntary communities; green design fosters and supports this strength. People who live in green communities stay longer, are more involved in community life, and are generally more likely to create a rich, vibrant community.

President Obama has committed the United States to cutting CO_2 by 83% by 2050. Given the prior administration's resistance to international efforts to address climate change, this new commitment is of deep importance. But the technical and institutional challenges of sharply reducing the production of GHGs, both within and outside the United States, are enormous. Reshaping our economy to dramatically cut climate-change gases is a staggering task; achieving it through energy efficiency and renewable energy would create very large employment, health, and societal benefits, and make the country economically more competitive.

Greening our built environment will require sustained federal support and broad industry engagement—but with such support and engagement, green design could become the design standard for almost all new construction and most retrofits by 2020. Our findings on the cost-effectiveness of green design, in both buildings and developments, demonstrate that "going green" would be a substantially lower-cost and lower-risk option than business as usual. In the green scenario outlined in part 4 of this volume, CO_2 from U.S. buildings would decline 20% by 2030, and by 60% by 2050—cost-effectively achieving one-third of the economy-wide target laid out by President Obama. The benefits of a national shift to green building—in terms of security, employment, and competitiveness—would also be large.

There have been extensive claims that cutting global warming could severely damage the economy, resulting in widespread job losses and damaging American competitiveness.

Our findings—that green design provides a highly cost-effective way of reducing CO_2 while creating jobs, strengthening property values, and increasing the health and resilience of communities—demonstrate that these claims are generally wrong. On reflection, it should not be surprising that cutting waste and improving design should be profitable.

Indeed, based on detailed analyses of 170 green buildings, we can state with confidence that greening buildings is generally cost-effective, whereas conventional development and design are likely to be risky and financially imprudent. A national commitment to green design and increased energy efficiency, along the lines of the green scenario described in this book, would create substantial national wealth—on the order of one trillion dollars. Given the reality and severity of climate change, a national shift to green design is both financially and environmentally wise.

Greening buildings is a cost-effective means of achieving relatively deep energy-efficiency gains and accelerating the deployment of renewable energy. Green design addresses both energy efficiency and health objectives; as a design approach, it requires integrated design, measurement, verification, and commissioning (to ensure that systems are installed properly). Green buildings typically achieve substantially greater efficiency gains than investments in energy efficiency alone. Greening also has a more visible brand than energy efficiency alone and is therefore a more motivating objective than energy efficiency alone. Greening is thus a powerful means of driving deep improvements in energy efficiency, in both new and existing buildings.

Green buildings are about 30 times more likely than conventional buildings to include on-site renewable energy (such as solar), or to buy power generated from renewable energy (such as wind, geothermal, or solar). A national shift to green buildings would drive rapid growth in demand and growth for renewable energy, driving the development of over 30 gigawatts of additional new renewable energy by 2020, and over 200 gigawatts by 2030. This reflects the very rapid projected growth of green buildings, the large demand for renewable energy in green buildings, and the fact that 75% of electricity is used by buildings. Green buildings in this scenario would thereby become one of the largest and most effective strategies for accelerating a national transition to clean energy.

The solution to the monumental problem of climate change will not come from one, or even several, huge centralized technological solutions, but will come mostly in small bites—and many of the cheapest solutions will be integrated into the hundreds of millions of buildings in the United States and globally. These solutions include energy efficiency, renewable energy, low-CO_2 materials, and smart choices about how we site our buildings in relation to each other—that is, whether our designs make driving the only way to get anywhere. Green design provides a tool for addressing all these opportunities in an integrated and therefore cost-effective way.

PART I

COSTS AND BENEFITS OF GREEN BUILDING

Perhaps because we spend the vast majority of our lives in buildings or traveling between them, we often overlook the scale of building energy use and the associated impact on climate change. For example, in a 2007 national survey of 1,000 home-owners, almost 75% said that they believed their homes had no adverse environmental impact.[1] The reality is quite different. According to the Energy Information Agency, residential and commercial buildings together consume 41% of the energy, including 74% of the electricity, used in the United States—a figure that does not include energy use in industrial office buildings. And of course, it also takes energy to make the materials necessary to construct and operate buildings (e.g., bricks, concrete, mechanical systems); to transport the materials; and to actually construct buildings. Despite widespread misperception, at least 45% of all energy used in the United States and Europe is consumed directly in buildings. The level of energy use and the resulting CO_2 emissions associated with buildings are almost as high as that from transportation and industry *combined*. Thus, the built environment provides a powerful and necessary lever for fundamentally changing our patterns of resource and energy use and responding to the grave reality of climate change.

1.1. Methodology

Over a 20-month period beginning in 2007, working with over 100 architects, developers, green building consultants, and building owners, we surveyed over 300 buildings and gathered detailed data on 170 green buildings, including the costs of going green; energy and water savings; and health, productivity, and other benefits.[2] We then synthesized the results of our survey with findings from other studies, to develop estimates of the present value of costs and benefits. The other studies took a number of forms and addressed a range of issues; they included large-scale building-performance surveys, health research, case studies, market studies, policy research, economic modeling efforts, and detailed analyses of the costs of green and nongreen buildings. To accompany this data and analysis, we solicited the perspectives of leading practitioners in the field: architects; academics; and corporate, nonprofit, and community leaders.

We sought examples of green buildings, primarily in the United States, that were either completed or under construction, and that were certified or anticipating certification through the U.S. Green Building Council's (USGBC's) Leadership in Energy and Environmental Design (LEED) rating system or other similar system.[3] Another criterion for inclusion was the availability of data on cost and performance that could be compared to data for a conventional version of the same building. Data were gathered directly from building owners, architects, and developers. (Appendix A describes the data-gathering methodology in detail.)

The 170 buildings for which the data sources (e.g., the architect or the developer) were able to provide information on the green premium—that is, the incremental cost of green building—make up the final data set used for cost-benefit modeling.[4] (Appendix C lists the major data points for each building.) To allow comparability of financial impacts over time, costs and benefits are expressed in terms of dollars per square foot ($/sf).

We looked at a wide range of building types, including schools, owner-occupied offices, offices built on spec, health care facilities, multifamily residential buildings, theaters, places of worship, college and university facilities, and laboratories (see figure 1.1). Because the buildings achieved LEED or equivalent certifications and a range of energy and water-use savings, we were able to evaluate the cost-effectiveness of differ-

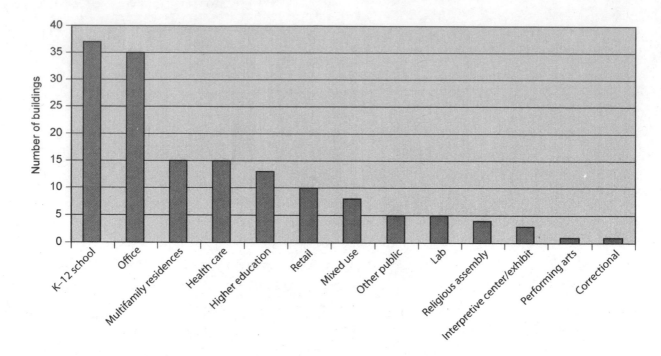

FIGURE 1.1 Building Types in the Data Set

ent levels of performance and benefits. The data set includes buildings in 33 states and eight countries, completed between 1998 and 2009 (see figure 1.2), with from 2,400 to 2 million square feet.[5]

BENEFIT MODELING

We developed net-present-value (NPV) and simple payback models to compare life-cycle benefits (including energy and water savings; emissions reductions; and increases in job creation, health, and productivity) with the initial cost of going green. Modeling requires assumptions, which are described in the relevant sections and in the appendixes. The general assumptions used in all of our present-value calculations are described in this section.

NPV calculations allow cost premiums to be compared with a subsequent stream of financial benefits. NPV represents the present value of an investment's discounted future benefits, minus any initial investment. Modeling NPV on a $/sf basis allowed us to compare initial building costs with a future stream of benefits.

Building benefits were calculated assuming a 20-year period—which tends to underestimate benefits, because 20 years is substantially shorter than the useful life of most buildings.

Present-value calculations of future benefits were based on a 7% discount rate. This rate is equal to or higher than the rate at which states, the federal government, and many corporations have historically borrowed money, and thus provides a reasonable basis for calculating the current value of future benefits. Unless otherwise noted, we assumed 2% annual inflation.

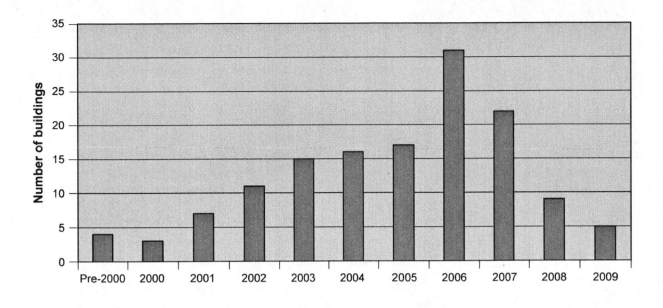

One of the primary challenges of any cost-benefit analysis is defining the baselines for the measurement of cost and performance. In the case of energy and water savings, the contacts for each building (typically the building architect or engineer) relied on industry standards to create a baseline for conventional buildings, against which green building savings could be measured. The building architect also provided the cost premiums, to allow a comparison between green building costs and the baseline. Standards and considerations that determined the selection of baselines and helped define our survey questions are addressed in detail in the other sections of the book and summarized in appendix E.

FIGURE 1.2 Buildings in the Data Set, by Year of Completion or Projected Completion

BASIS FOR BENEFIT MODELS

For over two-thirds of the buildings in our study, we were able to obtain information on energy and water savings, and for over one-third, data on construction-waste recycling and the use of recycled and local materials. Modeled benefits from energy and water savings and emissions reductions are based on building performance or attributes documented in this study, using appropriate assumptions. Estimated employment impacts are based on macroeconomic simulations run using data inputs from buildings in the data set.

To obtain credits, most LEED-certified buildings must undertake detailed modeling of energy and water savings and track the recycling of construction waste and the use of recycled and local materials. Impacts on health, productivity, and property values are relatively difficult to quantify, however, and are not required to be measured for green certification.[6] Information on health and productivity effects in the data set is therefore sparse. Additionally, a majority of our data sources were architects, who generally did not have access to information about ongoing effects on occupants. Thus, we used a synthesis of relevant literature and widely referenced models to quantify health, productivity, property value, and employment effects. We drew on a range of research, including surveys of occupants in green buildings, statistical analysis of real estate data from green buildings, and macroeconomic models of green building costs and energy expenditures.

PRIVATE VERSUS PUBLIC BENEFITS

The study models benefits that accrue in two distinct ways: (1) directly to building occupants and owners, and (2) indirectly to the surrounding communities and society at large. Both categories of benefits are described and presented in the models because both are substantial. Reductions in energy and water use and changes in operations and maintenance requirements commonly have direct financial consequences for building owners and occupants, as well as indirect impacts on society (e.g., decreased need for investment in expanding public water-treatment facilities).

Occupants experience direct health and productivity benefits, and employers and society experience indirect benefits. Reductions in emissions and storm-water flow, changes in employment brought about by new technologies, and changes in energy demand, for example, have financial consequences for state and local governments. The magnitude of these benefits is often hard to calculate precisely, but is generally significant. These benefits, therefore, should be material factors in developing green building projects, initiatives, regulations, requirements, and incentives.

LIMITATIONS

The broad approach taken here has some limitations that readers should bear in mind. Soliciting voluntary study participants and requiring that sources share certain types of data create a potential for bias in the selection of firms and projects. One might, for example, expect this data set to represent a generally more successful pool of projects than green buildings in general. In terms of cost premiums, however, it is not clear that the selection process would skew the data in only one direction; while some sources might want to share a cost-effective project, others appear eager to publicize buildings that showcase a large financial commitment to green goals.

Although the data set captures much of the diversity of the green building market in terms of geography, performance, and building type, the data set is not precisely representative of the actual national population of green buildings. For example, a comparison to the USGBC's records on certified and registered projects reveals that the buildings in the data set tend to be greener than average (e.g., greater reported reductions in energy use, a higher percentage of Platinum buildings). As a consequence, buildings in the data set would be expected to have slightly higher energy and water savings and reported green premiums than those associated with "average" green buildings built in the past decade (see appendix D). The bias toward greener buildings in this data set, which consists of buildings constructed over the past decade, coincides with the continuing trend toward greener buildings over time, suggesting that the data set provides a reasonable basis for anticipating the performance of new green buildings.

Finally, this book does not compare actual to projected performance; the primary focus is on the financial costs and benefits of green versus conventional buildings, given the best currently available information. In estimating long-term costs and benefits, we have used modeled costs and projected energy and water savings data where actual data were not available. For more information about measured performance of green buildings, see appendix I.

1.2. The Cost of Building Green

Question: How much does it cost to build a green building compared with a conventional building?

Evidence: The 170 (U.S.) buildings and 10 non-U.S. buildings in the data set reported green premiums ranging from 0% to 18%, with a median of 1.5%; the large majority reported premiums between 0% and 4%. Different approaches to researching the cost of going green yield similar results.

Bottom line: Most green buildings cost slightly more than similar conventional buildings to construct: the typical added cost of building a green building is $3/sf to $9/sf. Generally, the greener the building, the greater the cost premium, but all LEED levels can be achieved for minimal additional cost.

National and international surveys continue to reveal a widespread perception that green buildings cost substantially more to construct than conventional buildings. Recent surveys also find that concern over first costs remains the primary barrier to green building. For example, *Global Green Building Trends,* released in 2008, reports that of the over 700 construction professionals who responded to the survey, 80% cited "higher first costs" as an obstacle to green building.[7]

Some green architects and other experts believe that green buildings cost substantially more, while others emphatically believe that green design does not, or need not, cost more than conventional buildings. This discrepancy was evident during a single day of research for this book. Nick Berg, a partner in the development of Avanyu, a proposed mixed-use sustainable community and resort in Utah, reported his expectation for the green premium on the project: "This is our first green project and we don't know the premium yet, but anticipate with good planning no more than 20%."[8] In response to our inquiry on the green premium for two completed LEED apartment buildings, Michelle Rosenberger, of GGLO Architects, wrote as follows: "Shame on you for perpetuating this myth that green design costs more even if integrated properly. LEED certification does, but green design need not. I assume you are making that kind of distinction."[9]

A 2007 survey by the World Business Council for Sustainable Development found

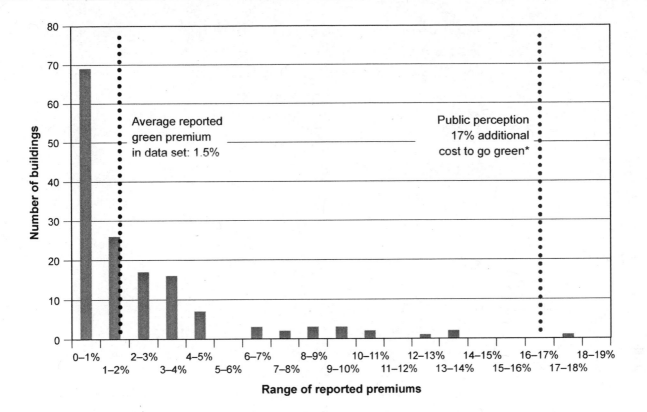

FIGURE 1.3 Green Premium for Buildings in the Data Set

Note: Public perception derived from World Business Council for Sustainable Development (WBCSD), "Energy Efficiency in Buildings: Business Realities and Opportunities," 2007 (www.wbcsd.org/DocRoot /lKDpFci8xSi63cZ5AGxQ/EEB -Facts-and-trends.pdf).

that business leaders believe that green building is, on average, 17% more expensive than conventional design.[10] Figure 1.3 illustrates this perception and compares it to the actual green premiums found in our study. Strikingly, the public also appears to underestimate the environmental impacts of buildings: the same international survey showed a public perception that buildings produce roughly 20% of CO_2 emissions, when in reality they account for almost half. And, as noted in the introduction to this volume, a recent survey of U.S. homeowners found that nearly three-quarters believe that their homes have no adverse environmental impacts.[11]

Developing green, walkable neighborhoods and communities is also commonly thought to be more expensive than conventional sprawl, but additional cost has not been clearly documented, and some green community-development techniques have been shown to result in substantial first-cost savings (see section 2.3, "Financial Impacts of Green Community Design").

DEFINING THE GREEN PREMIUM

In collecting data on the cost of green buildings compared with conventional design, we defined the green premium as the cost difference between green and nongreen (conventionally constructed) versions of the same building. All costs (e.g., construction, design, modeling, certification, etc.), except the cost of land, are included—an approach that permitted full comparability between green and conventional construction, and that took into account the characteristics of the particular project. Appendix F describes in greater detail some of the challenges of researching and defining the green premium.

FINDINGS: NEW CONSTRUCTION

The 170 U.S. buildings that make up the data set report cost differentials ranging from slight cost savings to 18% additional cost (see figure 1.4).[12] More than three-quarters of the buildings in the data set have green premiums between 0% and 4%; the largest concentration (69 buildings) is between 0% and 1%.[13] The median cost increase was 1.5%, and the mean cost increase was 2.8% before incentives.[14] These figures translate into a typical cost premium of about $3/sf to $9/sf.[15]

At the other end of the spectrum, nine green buildings in the data set reported a green premium of 10% or more; these include one Silver, four Gold, and four Platinum buildings. Thus, in this data set, there are more Platinum buildings with little or no green premium (0% to 2%) than with a large (10% or more) premium—suggesting that the cost premium depends more on the skill and experience of the design and construction team and on the choice of green strategies than on the level of greenness. Architects, engineers, contractors, and owners of green buildings almost universally report that early integration of green goals into the design process is crucial for achieving cost-effective designs.

Of the 170 buildings in the data set, 125 reported total project costs and premiums in

FIGURE 1.4 Green Premiums for Buildings in the Data Set
Each diamond represents one building.

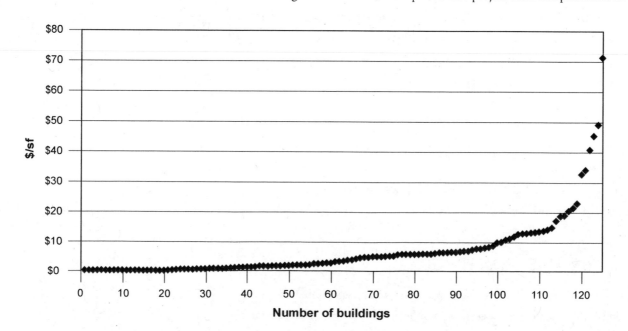

terms of dollars per square foot. As shown in figure 1.4, absolute green premiums ranged from $0/sf to $71/sf, with a median of $3.40/sf.

Our findings are in keeping with those of our previous studies, which have found that green buildings cost approximately 2% more to construct than conventional buildings. Two previous Capital E assessments, led by Greg Kats, used a similarly inclusive definition of green premium. "Costs and Financial Benefits of Green Buildings" and "Greening America's Schools: Costs and Benefits" surveyed 58 green offices and schools (40 of these buildings, with updated data, are included in the data set for the study on which this book is based). The surveys found that green buildings cost between 0% and 7% more than conventional buildings, with an average cost premium of slightly less than 2%. Many of the buildings included in these studies were early adopters of green building strategies.

Other researchers who have taken different approaches to assessing the green premium have arrived at a similar range of estimates. For example, a 2004 report by Steven Winter Associates for the U.S. General Services Administration evaluated the cost of green building for a model design of a federal courthouse and an office building.[16] When compared with a baseline model design, cost estimates for each LEED credit ranged from a slight cost reduction to 8% additional cost, depending on the LEED level pursued and the ability of a particular project to take advantage of low-cost LEED credits.[17]

In studies conducted in 2005 and 2007, Davis Langdon, an international building consulting firm, took a different approach, comparing per-square-foot costs for 83 LEED-seeking and 138 non-LEED-seeking buildings, including academic facilities, libraries, laboratories, community centers, and ambulatory care facilities.[18] Costs were normalized for location and date of construction, and comparisons were made by building type. There was no statistical difference between the cost of green and nongreen buildings. The authors of the study note that the study included examples of high- and low-cost buildings, both green and nongreen.[19]

Some or all of the additional up-front cost of green design is typically offset by savings resulting from the green elements. For example, improved insulation can reduce the size of the heating or cooling system; waterless urinals reduce plumbing requirements; and increased daylighting and views can decrease the required density of installed lighting. The model green school developed by the architectural firm OWP/P for the Chicago market includes a green roof that obviates the need for a water retention system (which is normally required by building codes), thereby decreasing capital costs and reducing the school's green premium to 1%.[20]

Although the green premium tends to be higher in buildings that incorporate more green elements, this is not universally true. For instance, many of the buildings in the data set with low (no more than 2%) or zero reported cost premiums are either Gold (29 buildings) or Platinum level (five buildings). Indeed, the data demonstrate that relatively green

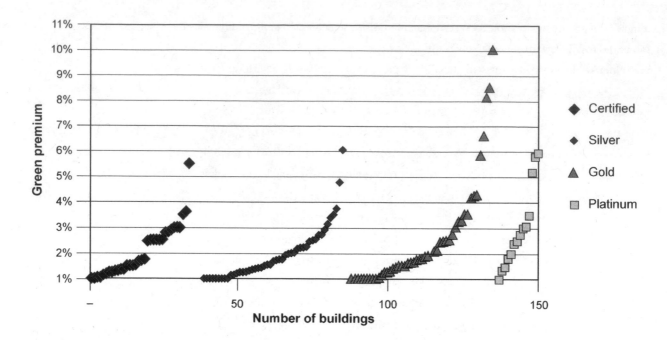

FIGURE 1.5 Green Premiums for Buildings in the Data Set, by LEED Level (Sorted by Increasing Premium)
Each shape represents one building.

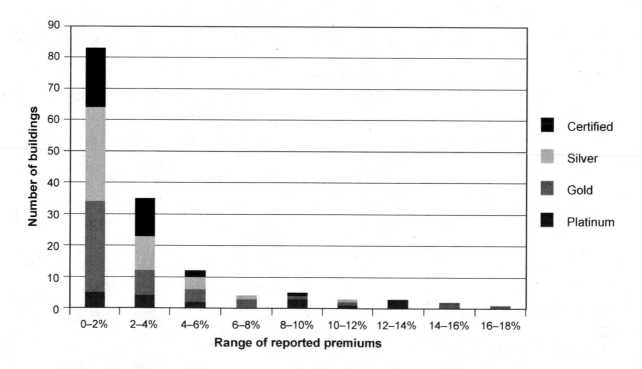

FIGURE 1.6 Green Premium Frequency for Buildings in the Data Set, by LEED Level

buildings can be built with virtually no cost premium, while some slightly green buildings can have a substantial cost premium. This pattern is illustrated in figures 1.5 and 1.6, which map cost premiums by LEED level.

FINDINGS: RENOVATIONS

While the majority of the buildings in the data set are new construction projects, 20 are full or partial renovations. These include gut rehabs of historic buildings, office fit-outs, and additions or partial renovations to existing structures. Green renovations are often thought to have higher green premiums, in part because features such as orientation and structural elements are typically not amenable to modification, which effectively limits greening options.

With a median of 1.9% and a mean of 3.9%, green premiums for the renovations were slightly higher than those for the data set as a whole. However, it is important to note that 25% of the renovations are LEED Platinum, whereas only 10% of the buildings in the full data set are Platinum. Thus, the premium for green renovations appears similar to the premium for new green buildings at the same level of greenness. This is an important finding, since it indicates that deep CO_2 reductions, derived from deep energy-use reductions in existing buildings, can be cost-effectively achieved through a national strategy of greening exisiting as well as new buildings. The USGBC's LEED for Existing Buildings standard provides a certification tailored to green retrofits, with a special emphasis on operations and maintenance of existing faciliites.

Although some green design choices are generally not available for renovations, most features—such as energy-efficient mechanical systems, water-efficient fixtures and landscaping, and green operations and maintenance practices—can be incorporated into renovations. Reusing (rather than demolishing) existing buildings is in itself an important resource-saving choice promoted by LEED standards.

1.3. Energy-Use Reductions

Questions: How much energy do green buildings use compared with conventional buildings? What is the value of energy savings in green buildings?

Evidence: Buildings in the data set reported a range of projected and actual reductions in energy use, from less than 10% to more than 100% (meaning that the building generates more power than it uses), with a median reduction of 34%. In terms of dollars per square foot, 60 buildings reported annual energy savings ranging from $0.10/sf to over $2/sf, with a median annual savings of $0.50/sf.

Bottom line: Based on the median savings from the data set and national data on baseline energy expenditures, the present value of 20 years of energy savings in a typical green building ranges from $4/sf to $16/sf, depending on building type and LEED level. Analysis of 18 buildings projected to reduce energy use by 50% or more demonstrates that advanced energy savings can be cost-effectively achieved with today's technology. For these buildings, the median green premium is 4%.

Energy savings are typically the most widely recognized and often the most bankable financial benefit associated with building green. Typical energy-saving enhancements include more efficient lighting, greater use of daylighting and sensors, more efficient heating and cooling systems, and better-insulated walls and roofs.

There are three types of energy savings in green buildings: (1) direct savings, which occur because more efficient buildings consume less energy; (2) indirect, economy-wide energy savings, which occur when drops in overall demand for energy drive down the overall market price for energy; and (3) "embodied energy" savings—that is, savings that result from reductions in the amount of energy used in materials and in building construction.[21] Reduced emissions—especially reductions in CO_2, the principal gas causing climate change—are increasingly recognized as a critical benefit of reduced energy use in green buildings.

In the discussion that follows, energy-savings estimates are presented by building type and by LEED level wherever possible. Given the wide range of energy intensities in green and nongreen buildings, and the wide range of variables that can affect energy use (e.g., region, building and systems design, occupancy, building management), these estimates should be viewed as typical of the data set; they are not intended as performance or cost predictions for specific buildings. In keeping with industry practice, reported reductions in energy use were largely based on computer models developed before construction. Actual energy use may vary significantly from the building projections. However, such projections for a portfolio of buildings have been shown to be relatively accurate and to provide a reasonable basis for cost-benefit comparisons. For a discussion of projected versus actual energy use in LEED buildings, see appendix I.

DIRECT ENERGY SAVINGS

When the 170 buildings in the data set were compared with conventionally designed buildings, the median reported energy-use reduction was 34%, with a mean of 35%.[22] Figure 1.7 shows the distribution of reported reductions in energy use for all building types.

Even within a single building type and region, green and nongreen buildings show a wide range of energy intensities, depending on factors such as building design, mechanical systems and appliances, operations and maintenance practices, and occupancy.[23] As shown in figure 1.8, energy savings in green buildings can also vary widely—from 0% to over 80%. In one instance—the Aldo Leopold Center—projected energy savings combined with on-site generation mean that the building is a net energy generator. Through efficient systems, including a ground source heat pump, daylighting, a highly insulating envelope, zoned heating and cooling, and on-site solar photovoltaic (PV) panels, this building is expected, on an annual basis, to produce more energy than it consumes. In contrast, it has been reported

FIGURE 1.7 Reported Reductions in Energy Use for Buildings in the Data Set

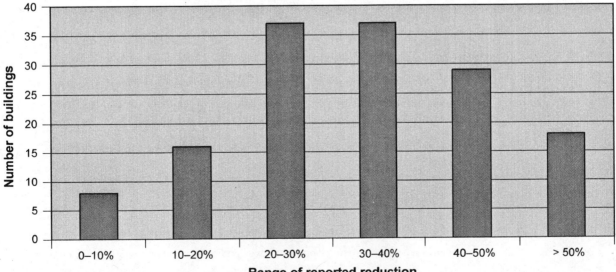

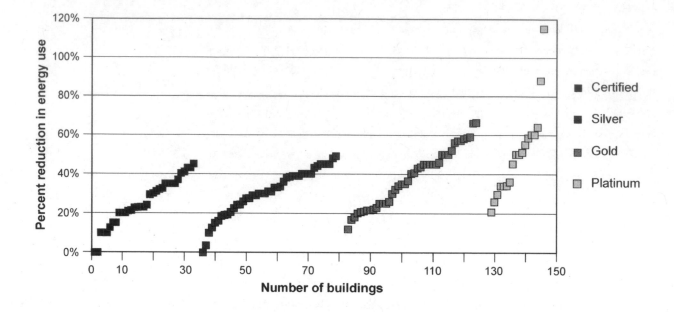

FIGURE 1.8 Reported Reductions in Energy Use by LEED Level

Each square represents one building.

that several recently constructed green buildings use more energy than national standard baselines (that is, in terms of energy use, the buildings yield "negative savings").[24]

In June 2007, the USGBC issued a new requirement that all new, LEED-certified buildings be designed to reduce energy use by at least 14% below the American Society of Heating, Refrigerating, and Air-Conditioning Engineers' (ASHRAE) 90.1 2004 standard; a 7% reduction is required for existing buildings.[25] (This requirement has since been updated; new buildings must now achieve a 10% reduction from the ASHRAE 90.1 2007 standard.) Although the buildings in the data set were registered for LEED before the requirement went into effect, we would expect energy reductions in future green buildings to increase in response to the gradually increasing stringency of the ASHRAE 90.1 standard referenced by LEED requirements. If all the buildings in the data set that reported energy savings of less than 14% had achieved 14% savings, there would have been a 1 percentage point increase in overall savings.

Projected energy savings generally increase with the level of greenness, and there is a range of projected savings at each LEED level (see figure 1.8). When compared with an ASHRAE 90.1 baseline building, LEED-certified buildings in the data set reported median savings of 23%; for Silver, the figure was 31%; for Gold, 40%; and for Platinum, 50%.

VALUING DIRECT ENERGY SAVINGS

Of the 170 buildings in the data set, 67 projected annual energy savings on a per-square-foot basis. (The remainder reported savings only as a percent reduction in energy use.) These 67 buildings included a similar distribution of building types and LEED levels as

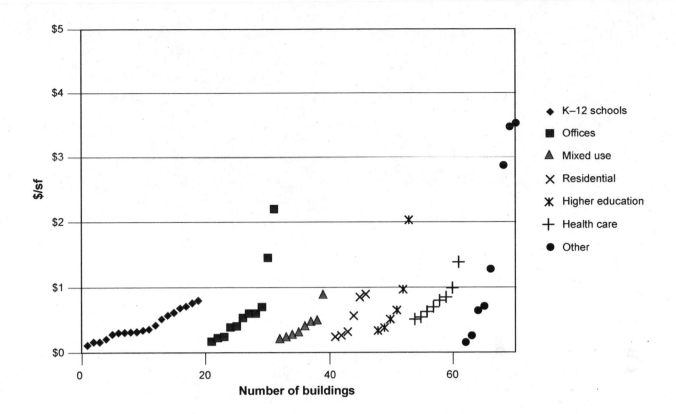

the larger data set. Figure 1.9 shows a scatter plot of these findings, sorted in increasing order by building type. Annual savings ranged from $0.20/sf to roughly $1/sf for the most common building types, with a median of $0.50.

Because of the small number of responses indicating dollars-per-square-foot savings for each building type, we used the median percentage of savings (34%) for the entire data set to estimate the savings for a typical green building for common building types. Baseline dollars-per-square-foot energy expenditures for each building type were drawn from the 2003 Commercial Building Energy Consumption Survey (CBECS).[26]

In the CBECS 2003 survey, average annual energy costs (including both electricity and on-site natural gas or fuel-oil expenditures) for the types of commercial buildings included in our study ranged from $0.65/sf (religious assembly) to $2.35/sf (health care), with an average of $1.46/sf.[27] As noted earlier, the 170 buildings in the data set project a median annual energy savings of 34% when compared with conventional design; this translates to typical annual savings, in 2008, of roughly $0.57/sf in green commercial buildings, ranging from $0.25/sf (religious assembly) to $0.99/sf (health care), depending on the building type. Assuming energy prices grow 3% faster than the annual inflation rate of 2%, and assuming a 7% annual discount on future energy savings, the present value of 20 years of energy savings ranges from $4/sf for religious-assembly buildings to $14/sf for health care buildings (see figure 1.10).

FIGURE 1.9 Annual Energy Savings by Building Type for 67 Buildings in the Data Set
Each square represents one building.

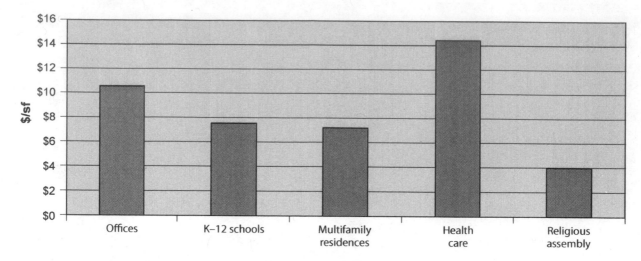

FIGURE 1.10 Present Value of 20 Years of Energy Savings by Building Type

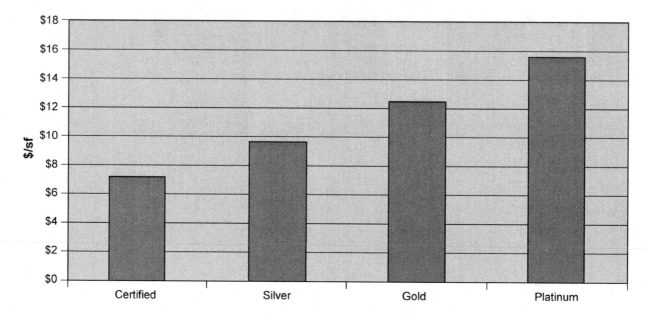

FIGURE 1.11 Present Value of 20 Years of Energy Savings in Green Offices, by LEED Level

(Energy-price assumptions and the sensitivity of energy-savings benefits will be discussed in detail later in this section.)

The value of energy savings also varies by LEED level. Taking typical energy expenditures in offices as an example, median reported reductions in energy use were used to estimate energy savings by LEED level. Using the assumptions described above for discount rate and increases in energy costs, the present value of 20 years of savings ranged from $7/sf for LEED-certified offices to more than $15/sf for LEED Platinum offices (see figure 1.11).[28]

The present value of energy savings from more efficient buildings depends heavily on future trends in energy prices—which are, of course, unknowable. In the first six months of 2008, average U.S. retail electricity prices for commercial buildings were roughly $0.10/kWh and had risen an average of 6% per year over the previous four years.[29] The average price of natural gas rose 7% annually over the same period.[30] A weighted average of these growth rates, assuming that 74% of spending on building energy is electricity and 26% is oil/natural gas, yields an annual 6% increase in energy prices between 2004 and 2008.[31]

Rapidly growing international demand for finite, nonrenewable energy resources; restrictions on and the rising cost of expanding generating and refining capacity; and restrictions on and the rising cost of expanding power distribution and transmission infrastructure strongly suggest future energy-price increases and price volatility. It is thus worth considering several possible trends in energy prices and the implied present value from energy savings. In the summary of estimated benefits presented in this book, we assume that energy prices will rise 5% per year. Under this assumption, the present value of 20 years of energy savings in green offices is $10/sf, three times the median reported green premium of $3.40/sf.

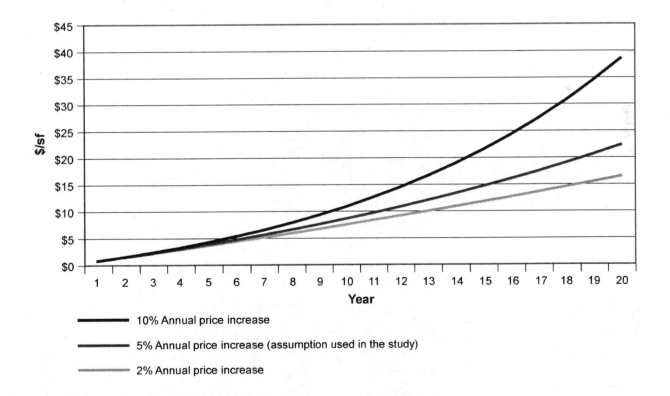

FIGURE 1.12 Cumulative Energy Savings in Green Offices: Sensitivity to Escalation in Energy Prices (Nominal)

If energy prices rise at only 2% per year (i.e., at the rate of inflation), then the present value of 20 years of energy savings in green offices is $8/sf, more than twice the typical green premium. However, if energy prices grow more rapidly, at 10% per year (i.e., 8% above inflation), the present value of the 34% lower energy use in green offices is worth about $17/sf, five times the median premium for greening buildings (see figure 1.12.)

A significant conclusion that can be drawn from these three illustrations is that even if energy prices stay flat (i.e., at 2%, the long-term inflation rate), discounted energy savings alone exceed the average green premium after five to eight years. If energy prices continue to rise at recent historical rates, then energy savings will be about three times the size of the cost premium, and will offset the cost premium in approximately five years. Thus, green buildings make financial sense from the perspective of energy savings alone. By reducing energy expenditures, building green provides a cost-effective hedge against the risk of future inflation and volatility in energy prices.[32]

Additional detail on the cost, methods, and issues in developing energy-savings estimates for green buildings can be found in appendix G, which compares reported additional expenditures on energy efficiency and renewable energy for 12 buildings in the data set to expenditures for conventional buildings. Appendix H provides a discussion of energy-savings data and baseline considerations, and appendix I summarizes the results of a recent USGBC/New Buildings Institute study that compares projected energy savings to actual utility bills for 121 LEED buildings.[33]

In addition to achieving reductions in the use of conventional energy through efficiency and the use of on-site renewable energy, 35% of LEED-certified buildings earn points for purchasing green power. These buildings agree to purchase at least 35% of their energy from green power programs for at least the first two years of operation; building owners or occupants purchase renewable energy directly from a utility, or purchase renewable-energy credits (RECs) that help fund new renewable-energy installations. Nationally, 1% of power customers participate in green power programs.[34] Thus, green building owners are currently approximately 35 times more likely than owners of nongreen buildings to purchase green power (at least for the first two years of operation). Green buildings constitute a significant and growing portion of the market for new renewable power, and are helping to reduce CO_2 emissions and drive expanded construction of renewable-energy installations.

INDIRECT ENERGY SAVINGS

As noted earlier, green buildings create indirect energy savings because substantial reductions in energy demand drive down energy prices across entire markets or regions. For an individual building, this secondary price impact is minuscule or nonexistent—but statewide or nationally, the secondary impact of reduced energy consumption can be substantial, and should therefore be a material factor for policy makers to consider in implementing green building and energy-efficiency programs across cities, states, and regions.[35]

Reductions in energy use help avert the need for new energy sources and new transmission and distribution capacity. Because these new assets are typically significantly more expensive than efficiency improvements, reducing the growth in energy demand through green buildings can create large savings. A recent study by McKinsey & Company suggested that a $160 billion investment in energy efficiency in buildings and appliances and a $90 billion investment in industrial efficiency across the United States through 2030 could result in $300 billion savings from avoiding investment in new power generation.[36]

Efficiency-driven reductions in demand can have significant impact on price. A 2005 report from the Lawrence Berkeley National Laboratory (LBNL) reviewed 19 national and state analyses of the impact of reductions in natural gas demand on wellhead price (the price of the natural gas commodity, excluding transportation and distribution costs). These studies, conducted by the American Council for an Energy-Efficient Economy, the Tellus Institute, the Union of Concerned Scientists, the Energy Information Administration, and others, found that a 1% reduction in demand resulted in price reductions of 0.8% to 2%.[37] A 2004 Platts Research & Consulting review of nine separate studies determined that a 1% drop in demand could drive a 0.75% to 2.5% reduction in long-term wellhead prices.[38]

These studies indicate that a reduction in natural gas consumption (and savings in energy costs) could drive a reduction in long-term natural gas prices equal to 100% to 200% of the direct savings from the reduction. Electricity prices are less volatile than natural gas prices, but still have been shown to respond to changes in demand. A 2004 Massachusetts state report analyzing the impacts of statewide electricity- and gas-efficiency programs concluded that during a year in which direct savings were estimated at $21.5 million, indirect savings (from lower overall energy prices caused by lower energy demand) equaled $19.4 million—that is, indirect savings were 90% of the direct savings from the efficiency programs.[39] Based on research findings, some of which are summarized above, the indirect savings attributable to a drop in demand, and the consequent reductions in energy costs brought about by a broad shift to green building design—though hard to estimate exactly— is material and should be included in benefits estimates. In this book, the value of indirect savings is, probably conservatively, assumed to be equal to 25% of the direct savings.

This indirect reduction in energy costs derived from energy efficiency has a present value of approximately $2/sf. (To keep this estimate simple, and to reflect the high level of imprecision, this saving estimate is applied across the portfolio of buildings.) Thus, the total direct and indirect present value of energy savings over 20 years from green building is estimated to be between $6/sf and $18/sf, depending on building type, energy intensity, and the level of savings achieved.

EMBODIED ENERGY SAVINGS

A significant amount of energy is consumed in the extraction, manufacture, and transport of building materials. Basic structural elements such as steel, concrete, and drywall are among the most significant industrial users of energy. Nationally, up to 16% of the total energy consumed by buildings is estimated to be embodied in materials, construction, and renovation.[40] The production of cement alone consumes 1% of the energy used in the United States, and has been estimated to be responsible for 2% of U.S. CO_2 emissions and between 5% and 8% of global CO_2 emissions.[41] A 2006 case study of a typical U.S. office building in Minnesota estimated that the total energy embodied in materials, construction, and maintenance was roughly 1.2 million Btu/sf, equivalent to almost ten years of the operating energy used in the building.[42] A series of case studies on office buildings in Australia found embodied energy from 950,000 Btu/sf to over 1.6 million Btu/sf.[43] As building operations become more efficient, a larger portion of building energy use and CO_2 emissions comes from materials and construction.

Green buildings, such as LEED-certified buildings, reduce embodied energy through the use of recycled, reused, and locally extracted and manufactured materials. Many recycled materials—including recycled metals, and cement that has high recycled fly-ash content or that is created using new, low-CO_2 processes[44]—require significantly less energy to produce than virgin materials. Green buildings in the data set used an average of 23% recycled materials and 35% locally produced materials, according to the definitions and calculation methods used in LEED. (Materials extracted and manufactured within 500 miles of the site are considered local; with respect to building materials, 100% of post-consumer and 50% of pre-consumer recycled content is counted toward percent recycled.)[45] Of LEED for New Construction–certified buildings, 9% earn an additional point by using at least 5% reused or salvaged building materials in construction.[46] Savings in embodied energy and embodied CO_2 emissions are not estimated for individual buildings in the data set, but are included in the green building scenarios in part 4.

VALUING CO_2 REDUCTIONS

Emissions reductions in green buildings are an essential component of a national CO_2 reduction strategy, and would provide substantial financial and nonfinancial benefits. One way to assess the financial value of CO_2 reductions is to examine the price of CO_2 reductions through global carbon taxes or markets. Although there is currently no U.S. carbon tax or national mandatory market for carbon emissions, regional carbon-trading markets do exist.[47] Prices on the European carbon market have fluctuated since its inception, and were roughly $20/ton as of mid-2009.[48] The value of CO_2 reductions relative to the value of energy savings is highly dependent on whether—and at what price—future reductions can be sold.

Figure 1.13 shows the present value of 20 years of CO_2 reduction and direct energy and water savings for a typical green office, at four possible market prices for CO_2:

- At $5/ton, CO_2 reductions are worth less than $0.50/sf over 20 years. This price is close to 2008 prices on the Chicago Climate Exchange.
- At $10/ton, CO_2 reductions are valued at just under $1/sf over 20 years.
- At $20/ton, CO_2 reductions are estimated at $1 to $2/sf. This price is close to the mid-2009 European trading price, and to the Obama administration's reported target starting price for a national CO_2 market.[49]
- At $50/ton, the value of CO_2 reductions is more than $3.50/sf over 20 years, which would pay back green premiums on typical green offices. This is close to peak pre-recession prices on the European CO_2 market.

In each scenario, the prices for CO_2 reductions are assumed to remain flat (that is, to rise at an annual inflation rate of 2%). Although there is much uncertainty about future carbon prices—and uncertainty about who will be able to claim CO_2

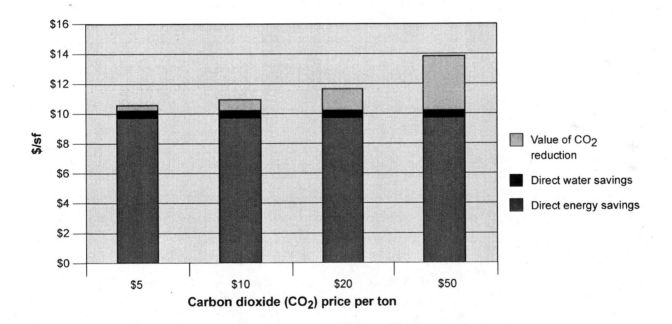

FIGURE 1.13 Present Value of 20 Years of Carbon Dioxide Reductions in Green Offices
Note: Estimated carbon dioxide (CO_2) reduction based on average office energy use for 2000–2003, as reported in Energy Information Administration, "Commercial Building Energy Consumption Survey," 2003 (www.eia.doe.gov/emeu/cbecs/cbecs2003/detailed_tables_2003/2003set9/2003html/c12.html). Average CO_2 emissions per unit of energy generation as reported in M. Deru and P. Torcellini, "Source Energy and Emissions Factors for Energy Use in Buildings," National Renewable Energy Laboratory, June 2007 (www.nrel.gov/docs /fy07osti/38617.pdf); U.S. Environmental Protection Agency, "Updated State-Level Greenhouse Gas Emission Coefficients for Electricity Generation, 1998–2000" (www.eia.doe.gov/pub/oiaf/1605/cdrom/pdf/e-supdoc.pdf).

reduction credits—it is widely expected that a carbon cap will be adopted in the United States, which will add value to energy-use reductions in green and energy-efficient buildings.[50]

Because continued global warming will have enormous negative economic impacts, the range of CO_2 prices discussed here do not reflect the full social value of reducing greenhouse-gas (GHG) emissions in green buildings. Estimates of this "social cost of carbon" vary widely: although a review of 100 estimates conducted by the Intergovernmental Panel on Climate Change (IGPCC) found an average of $12/ton, the IGPCC notes that the estimates are likely to significantly underestimate full social costs.[51] According to a 2007 economic analysis of the impacts of climate change by the U.K. Department of Treasury, the global cost of unmitigated ("business as usual") climate change is likely to be the equivalent of reducing global economic output by 5% to 20%.[52]

In modeling the financial benefits of green building in this book, we assumed that CO_2 emissions reductions had an average value of $15/ton to $20/ton (about the mid-2009 price of CO_2 on the European market), and that the price of CO_2 will go up with inflation. Based on these assumptions, we estimate that over 20 years, the present financial value of CO_2 emissions reductions is $1/sf to $2/sf, only partially reflecting the full social value of slowing growth in global climate change.

OTHER EMISSIONS REDUCTIONS

Reductions in non-CO_2 emissions from green buildings have additional health and ecological benefits. Sulfur dioxide (SO_2), a major pollutant associated with power generation, is the principal cause of acid rain, causes respiratory distress, can aggravate existing heart and lung conditions, and can react with other materials in the atmosphere to form particulates.[53] Nitrogen oxides (NOx) are involved in the formation of smog, which impairs lung function and can damage lung tissue, especially among children, elderly people, and people with respiratory conditions such as asthma.[54] Particulate matter has been implicated in a number of respiratory problems and may be responsible, according to some estimates, for as many as 60,000 premature deaths in the United States each year.[55] Heavy metals, such as mercury, have been linked to neurological damage and are especially harmful to children and pregnant women.[56]

The value of SO_2 and NOx emissions reductions can be estimated on the basis of market prices for such reductions on regulated allowance markets. However, emissions market prices can substantially undervalue the full societal impact of emissions reductions, including health and ecological effects.[57] Emissions of NOx, sulfur oxides (SOx), mercury, and heavy metals per unit of energy produced vary dramatically, depending on how the energy is generated. This book does not present national estimates for the value of NOx, SOx, particulate, or heavy-metal emissions reductions from green building, but such estimates can be made for a specific building or portfolio—based on the emissions intensity of local energy generation, and using existing tools for estimating and valuing the health impacts of reductions in emissions.[58]

1.4. Advanced Energy-Use Reductions

Green buildings generally achieve energy savings of 20% to 50% through measures such as proper building orientation; cool roofs; highly insulated walls and roofs; daylight harvesting; and the use of efficient lighting, heating, cooling, hot-water, and ventilation systems. Energy savings that are greater than 50% commonly require new or innovative technologies or design strategies. Carbon-neutral buildings—that is, buildings that generate as much or more energy than they consume annually—combine deep improvements in energy efficiency with substantial on-site renewable-energy generation. Reducing the total carbon impact of buildings will also require the use of materials with lower embodied energy. The following analysis shows that cutting energy use by half in green buildings can already be cost-effective—an important conclusion, given the plethora of state and national short-term targets for deep cuts in emissions and energy use.

EXAMPLES FROM THE DATA SET

Table 1.1 lists the 18 buildings in the data set that projected energy-use reductions of 50% or more, when compared with the ASHRAE 90.1 baseline. Energy-efficiency savings

TABLE 1.1 BUILDINGS IN THE DATA SET WITH MORE THAN 50% REDUCTIONS IN ENERGY USE

Reference number	Building name	Building type	Year completed	Energy savings (%)	Green premium (%)	Energy provided by on-site renewables (%)[1]
1	Dell Children's Medical Center of Central Texas	Health care	2007	50	4.0	0
2	Melink Headquarters	Office	2005	50	9.6	5
3	One Bryant Park (Bank of America building)	Office *(high rise)*	2008	50	2.0	0
4	Banner Bank Building	Office *(high rise)*	2006	51	0.7	0

TABLE 1.1 (CONTINUED)

Reference number	Building name	Building type	Year completed	Energy savings (%)	Green premium (%)	Energy provided by on-site renewables (%)[1]
5	Twin Lakes Elementary	K–12 school	2007	52	0.0	0
6	Heifer International Center	Office	2005	55	13.8	0
7	Bronx Zoo Lion House	Zoo exhibit *(plants and animals from Madagascar)*	2006	56	4.4	Not available (N.a.)
8	C. K. Choi Building	Higher education	1996	57	0.0	0
9	The Henry	Mixed use *(condos and above ground-floor retail)*	2004	57	1.0	0
10	Langdon Woods Residence Hall	Higher education	2006	58	2.2	0
11	Toyota Motor Sales South Campus Headquarters	Office	2003	59	0.0	20
12	Clearview Elementary	K–12 school	2002	59	1.3	0
13	Lewis and Clark State Office Building	Office	2005	60	4.0	2.5
14	Sidwell Friends Middle School	K–12 school	2006	60	9.6	5
15	Robert Redford Building, Natural Resources Defense Council, Santa Monica Office	Office	2003	64	13.6	20
16	Sunrise Yard *(maintenance and office building for the New York City Department of Transportation)*	Office	2008	66	2.9	0
17	Kirsch Center for Environmental Studies, De Anza College	Higher education	2006	88	5.5	N.a.
18	Aldo Leopold Center	Office	2007	113[2]	12.5	113
	Median			**58**	**4**	

Notes:
1 Energy generated by on-site renewables is included in the figures for energy-use reductions.
2 On an annual basis, the center is projected to produce 13% more on-site energy than it uses.

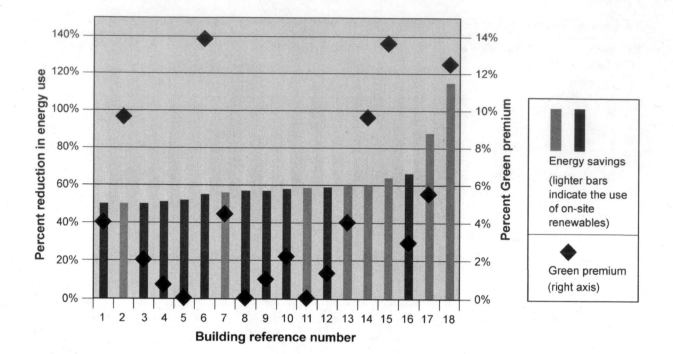

FIGURE 1.14 Advanced Energy Savings and Green Premiums for Buildings in the Data Set
Note: Each bar represents a building shown in table 1.1, by reference number.

range from 50% to over 80%. Reported green premiums for these buildings range from 0% to 14%, with a median green premium of 4%—higher than the 1.5% median green premium for the data set as a whole.[59]

Figure 1.14 shows the energy savings and green premiums for each of the 18 buildings. Eight of the buildings (indicated in the figure by yellow bars) generate energy from on-site renewables (this is reflected in the calculations of energy savings). Seven of the buildings obtain on-site renewable energy through a PV array; the Bronx Zoo Lion House (#7) has a 200-kW fuel cell installed in the basement. In all cases, energy-efficiency measures were used to significantly reduce the building energy load. For instance, at the Aldo Leopold Center, which had the goal of becoming a carbon-neutral building, energy-efficiency measures were used to reduce building energy use by 58%, and a sufficiently large PV array was installed to exceed the remaining load on an annual net basis.

SPOTLIGHT ON TECHNOLOGY

The foundation of improved energy performance in green buildings is energy-efficient design, construction, and operation, including proper building orientation; right-sizing and control of mechanical systems; efficient lighting, heating, cooling, and hot-water systems; insulation; and high-performance windows. The 18 buildings in the data set that reduced conventional energy use by 50% or more used a range of technologies, including ground source heat pumps and on-site solar PV. One of the buildings used Sage electrochromic glazing, an emerging building technology that is also described below.

Of course, innovation in green building technology is a moving target. Many energy-efficiency strategies, including low-emissivity (low-e) windows and reflective roofs, are rapidly becoming standard in green and nongreen buildings alike. Adoption of ground source heat pumps (GSHPs) has accelerated to around 75,000 new U.S. installations per year. Venture-capital investment is helping to drive rapid clean-energy innovation and reduction in costs. PV prices fell 40% between early 2008 and mid-2009. As production volume ramps up and prices drop—with the first commercial-volume production, in 2011—wide adoption of electrochromic glass is expected. According to Dave Deppen, the architect of the LEED-Platinum-rated Kirsch Center for Environmental Studies at De Anza College, learning about green features such as photosensors and low-e windows is the crucial step for most clients: "Once they know about them [green building features]—they become baseline."

Ground Source Heat Pumps

Ground source heat pumps, sometimes referred to as geothermal heat pumps, use the relatively constant temperature of the earth to reduce the amount of energy used for heating and cooling in a building. At depths of six feet or greater, the earth maintains a relatively constant temperature of typically between 50° and 60°F year-round. In a GSHP system, water or a refrigerant is run through vertical or horizontal underground pipes, and is heated or cooled to approximately 55°F by the ambient ground temperature; this reduces the amount of gas or electricity needed for heating and cooling. The GSHP system at the Aldo Leopold Center, for instance, consists of tubes in 19 holes, drilled 220 feet into the ground.

Used in one-third of the buildings in the data set that achieved greater than 50% energy savings, GSHPs are several times as efficient as conventional electric heating and cooling equipment, typically saving 25% to 50% of energy use.[60] GSHPs can be integrated into a variety of distribution systems, including under-floor air systems and hydronic cooling and heating systems. The largest additional expense of installing a GSHP system is drilling the holes for the underground pipes; however, in areas with sufficient available land, this expense can be reduced by installing horizontal pipes. A GSHP can be twice the cost of a conventional heating and cooling system, and the payback occurs within five to ten years in most cases.[61] Moreover, operating and maintenance costs for GSHPs are typically lower than for conventional heating and cooling systems. Other heat pump technologies, including air-source, water-source, and integrated heat pump water heaters, also show promise for reducing energy use by drawing on relatively stable or moderate temperature sinks in ambient air or water.

Electrochromic Glazing

Conventional low-e glass, which is used in many green buildings to reduce heat loss through windows, requires shades or blinds to adjust incoming light throughout a day, limiting views and adding maintenance costs. Electrochromic glazing is an emerging technology that allows windows to dynamically tint up or down, darkening to block solar heat gain when building cooling is desired and lightening to allow solar heat gain and natural light when it is cool and overcast.[62] In a study of the energy-savings potential of electrochromic glass used with daylighting controls, the LBNL estimated a 19% reduction in peak cooling loads when compared with a building with blinds and daylighting controls, and a 44% reduction in energy used for lighting when compared with a typical building with blinds and no daylighting controls.[63] Occupants also reported greater satisfaction with the controllability of electrochromic glass, when compared with static windows. The Twin Lakes Elementary School, in the Elk River Area School District in Minnesota (#5 in table 1.1), recently installed electrochromic glass in one of two adjacent science classrooms, as an interactive experiment in energy use and occupant satisfaction with the technology.[64] Electrochromic glass remains significantly more expensive than conventionally efficient options, but as its cost drops, it is beginning to be specified in a variety of commercial and residential buildings.

Solar Photovoltaics

Seven out of the 18 buildings that achieved advanced energy savings used on-site solar PV panels, indicating the critical role that PV will play in reducing carbon emissions from buildings. PV panels, which allow direct conversion of sunlight into electricity and thereby decrease the quantity of electricity that must be purchased from the grid, can be installed on roofs or building grounds, or can take the form of building-integrated photovoltaics. Electricity from PV produces no CO_2 emissions (apart from those that are embodied in the manufacture and transport of system components), and is not subject to future price fluctuations, since the fuel is sunlight. Fourteen percent of LEED-NC-certified buildings earn a LEED point for installing on-site PV to provide at least 2.5% of building energy needs. Prices for PV have dropped by a third since early 2008, but generally remain substantially higher than grid electricity on a dollars-per-kilowatt basis. However, PV typically reduces grid power use during peak hours, making PV more cost-effective.

Utilities and local, state, and federal government offer a range of incentives to partially offset the initial high cost of PV installations.[65] While incentives did not appear to make a significant difference in the cost-effectiveness of green buildings as a whole in this study, several buildings reported receiving incentives without which PV installations would not have been financially feasible.

COST-EFFECTIVENESS

Are these advanced energy-saving strategies cost-effective? The wide range of reported premiums for the 18 buildings that achieved at least a 50% reduction in energy use suggests that although additional expense is required, well-integrated design and careful choice of technologies can minimize the cost premium; in fact, seven of the 18 buildings reported cost premiums of 2% or less. The cost-effectiveness of technologies such as GSHPs and solar PV depends, in part, on local utility rates, subsidies, and building energy use. In areas with high rates for peak electricity use, on-site generation or load-shifting and peak-reduction technologies can be cost-effective. All but one of the buildings with PV reported overall green premiums above the study average, including the cost of PV and any applicable incentives.

Seven of the 18 buildings projecting more than 50% reductions in energy use had data available on dollars-per-square-foot energy savings, based on local energy rates and expected usage. Figure 1.15 shows the present value of 20 years of energy savings in these buildings, and the simple payback—that is, the number of years it would take to recoup the entire green premium through energy savings alone. In terms of dollars per square foot, the greatest energy savings are achieved in building types that have the highest energy use. In five of the seven buildings, energy savings alone pay back the green premium in less than five years. Energy-related costs generally make up the majority of the premium, especially for buildings using emerging technologies and on-site renewable-energy generation. In effect, more cost-effective technologies, such as insulation or efficient lighting, can help subsidize the cost of cutting-edge technologies, keeping overall payback periods reasonable while making it possible to achieve greater savings.

FIGURE 1.15 Simple Payback and Present Value of 20 Years of Energy Savings in Buildings with Advanced Energy-Use Reductions

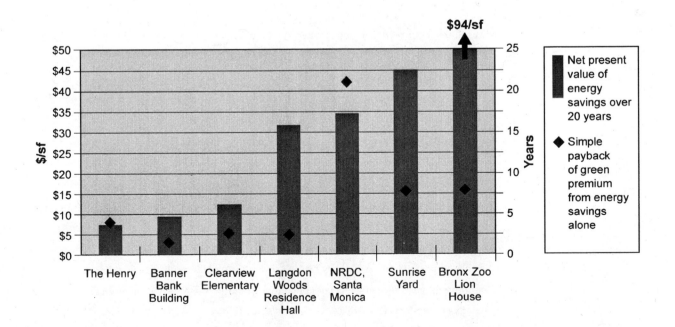

Robert F. Fox Jr., *Partner, Cook + Fox Architects*

Why would anyone want to build yesterday's building? With a skilled team of architect, engineer, owner, and builder, the additional costs of building green are minimal. We've found that the harder we work and the more creative we are in integrating architectural and engineering principles, the less it costs to build a green; high-performance building.

I can't tell you that a Platinum building will cost less than a LEED-certified building. But anyone who engages his or her team in an intelligent way and then makes sure the team members work hard together can build a very green building without much additional cost—certainly less than 3%.

For example, at One Bryant Park, Bank of America's new 2-million-square-foot, $1.3 billion headquarters in Manhattan, we knew we wanted to make our own power on site. Our team found that the most economical system would be a gas-fired turbine sized to provide 5 megawatts of power, about 67% of the building's annual energy needs. At night, however, the turbine would make more energy than we would need, so our engineers came up with the idea of making ice at night, which could be used to supplement the air-conditioning system during the day, significantly reducing energy use for cooling. The payback for the on-site cogeneration and ice-storage system is a few years.

We start almost all of our projects with a charrette—bringing together the entire team at the outset really gets the creative juices flowing. We never know where the good ideas will come from—the engineer, architect, contractor, or owner. When you are part of a team that is challenging itself to be creative, good things start to happen. It takes a bit more time, but we don't charge extra for it; it's what all architects and engineers should be doing.

The workplace is where people spend the majority of their day, so when the indoor environment is filled with daylight and fresh air, and provides a comfortable temperature and contact with nature (a concept known as biophilia), and when carpets and furnishings don't give off noxious emissions, it all adds up to a higher-quality work environment. With these strategies, a 1% improvement in productivity—the equivalent of five minutes a day—is certainly achievable. At the Bank of America Tower we think we'll get a 10% improvement, which is worth an estimated $100 million a year in boosted productivity. The cost of green features on the building is $20 to $30 million, so if we achieve a 1% improvement in productivity, $10 million in annual productivity benefits means a two- to three-year payback for all of the green features. If we get a 3% improvement in productivity, the payback is one year. The people making decisions at Bank of America are thrilled.

When we met with the chairman of Bank of America, Ken Lewis, he told us he wanted a building that would be an icon for New York City, one that would help attract and retain the best employees. What better way to attract and retain employees than to create an exceptionally healthy working environment? Like other financial institutions, Bank of America wants to hire the overachievers, the best talent out there. We'll be able to help them do that, and show them productivity impacts as well.

Reduced operating costs are another benefit of green buildings. Equipment that's commissioned works better, last longer, and require less maintenance.

Along with Bank of America, the green features of the building are attracting high-quality tenants who want to show their clients that they practice what they preach. Our hope is that once people realize that green building is the right thing to do, that it works and can even be profitable, more owners and architects will follow the example.

These examples (See Figure 1.14) demonstrate that energy efficiency can reduce energy use cost-effectively today by 50%. Adding on-site renewable energy generation would allow most of these very efficient buldings to achieve close to a 75% reduction in CO_2 from operations. California and Massachusetts have set state-wide goals that by 2020 new homes will be zero net energy. Caifornia Public Utility Commissioner Dian Greunich observes, "we have the technology and we are developing the state and utility incentives and programs to drive this as a state wide policy." The European Parliament set a 2019 deadline for all new homes built in Europe to be zero-net energy. Expanded public funding and mandates and accelerating venture capital investment in improving performance, reducing the price and ramping up volume of efficiency and renewable energy technologies means that scale construction of zero net energy/carbon buildings by 2020 is a realistic policy objective.

1.5. Water-Related Savings

Questions: How much water do green buildings use compared with conventional buildings? What is the value of water savings in green buildings?

Evidence: Of the 170 buildings in the data set, 119 reported or projected reductions in indoor potable water use when compared with conventional buildings; reductions ranged from 0% to more than 80%, with a median of 39%. Water savings generally increase with LEED level.

Bottom line: The present value of 20 years of water savings in typical green buildings ranges from $.50/sf to $2/sf, depending on building type and LEED level. Additional benefits of water-use reduction include decreased need for public water infrastructure and reduced water pollution.

Green buildings reduce the use of potable water through a variety of strategies, including efficient plumbing fixtures, rainwater harvesting, on-site wastewater treatment and recycling, and the use of native or drought-tolerant plants in landscaping. Water use in a building is often broken down into three categories: indoor fixture use (e.g., sinks, toilets, and showers), outdoor irrigation use, and process use (e.g., cooling, hospital equipment, laundry, or industrial use). Direct financial benefits of water-saving strategies include reduced charges for the provision of water and the treatment of wastewater; indirect benefits include reductions in infrastructure costs, reductions in the energy used to convey and treat water, and reductions in the energy used to pump and heat water. Although water costs usually make up a small part of the operating budget of buildings, water-use reduction strategies can be quite cost-effective, especially in areas with high water rates. Water conservation in buildings is generally a much lower-cost alternative than the upgrading or expansion of the water infrastructure system.

Many green buildings reduce storm-water runoff through strategies such as rainwater harvesting; storm-water reduction yields related benefits in areas where wastewater and storm-water systems are connected. Reductions in water use and storm-water runoff have the additional benefits of relieving stress on water systems and reducing the costly ecological and health impacts of storm-water and sewage overflows.

In green developments, significant additional reductions in water use and storm-water runoff are achieved through clustering, bioswales, pervious pavement, and habitat restoration. Conservation development strategies to reduce storm-water runoff and increase open space have resulted in up-front development cost savings of $12,000 per home (see section 2.3, "Financial Impacts of Green Community Design").

REPORTED WATER-USE REDUCTIONS AND STRATEGIES

The buildings in the data set reported a range of reductions in potable water use, largely based on analysis that takes into account expected occupancy and flow rates through faucets, showers, toilets, and urinals. The baseline flow rates used in these calculations are set by the Energy Policy Act of 1992 (EPAct 1992). Since few nongreen buildings exceed EPAct 1992 requirements, water-use reductions benchmarked against EPAct requirements provide a reasonable means of comparing water use in green and nongreen buildings.[66] If rainwater or recycled wastewater is used in place of potable water for irrigation or toilet flushing, that quantity is included in the reported water-use reduction. In the 120 buildings that were able to provide data on water-use reduction, reductions ranged from 0% to 94%, with a median of 39%.

Generally, greener buildings show greater savings, though there is a wide range of reported savings within each LEED level. Median reported water savings are 21% for Certified buildings, 36% for Silver, 39% for Gold, and 55% for Platinum. Figure 1.16 shows water-use reductions, by building type, for some of the major categories in the data set. The wide range of water savings in each category suggests that the potential for water savings does not appear to be limited by building type. (For a discussion of water savings in green health care facilities, see section 1.8, "Green Health Care: Assessing Costs and Benefits.")

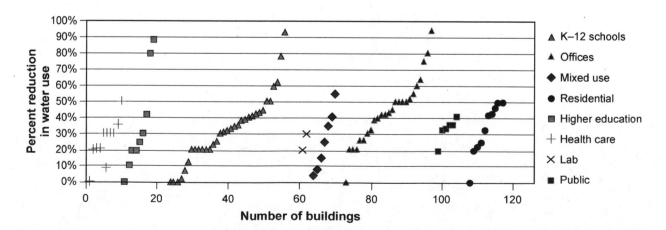

FIGURE 1.16 Water-Use Reductions by Building Type

Like energy-savings projections, water-savings projections are approximate measures of actual savings. A 2006 review of the post-occupancy performance of 11 green buildings in the Northwest included seven with records of actual versus projected water use. In these seven projects, which included offices and apartment buildings, median actual water use was found to be 15% higher than projected use.[67] The differences between actual and projected use probably stem from inaccuracies in the assumptions used to model user behavior and building occupancy.[68]

Many buildings use additional water for irrigation, cooling, and other process uses, which is not included in the reported water-use reductions just discussed. Depending on building type, location, and occupancy, these other water uses can be smaller, or much larger than fixture uses. For example, roughly two-thirds of the water used in hospitals is for process uses not regulated by EPAct.[69] The Green Guide for Health Care and LEED for Health Care award points for strategies to reduce process water usage (see section 1.8, "Green Health Care: Assessing Costs and Benefits").

Roughly 60% of LEED-NC-certified buildings use no potable water for irrigation.[70] Estimating the magnitude of green building reductions in the use of water for irrigation is difficult, however, given the lack of national standards or surveys for irrigation water use in buildings.

COST OF WATER-SAVING STRATEGIES

Water-saving fixtures can reduce indoor potable water use by 20% to 60%, and their initial costs are comparable to those of conventional options. Rainwater harvesting and on-site treatment and recycling of wastewater usually add expense to a project.[71] On the other hand, waterless urinals, composting toilets, and reduced irrigation can eliminate the need for some water lines, reducing maintenance costs and providing some construction-cost savings. Capturing and using or infiltrating storm water on site can also reduce the need for investments in storm-water drains and culverts, depending on local zoning regulations. Appendix J includes a cost breakdown of water-conservation measures in six green buildings and illustrates the initial financial impact of water-use reduction strategies.

PERSPECTIVE: WATER-SAVING STRATEGIES: OREGON HEALTH SCIENCES UNIVERSITY CENTER FOR HEALTH AND HEALING

The Oregon Health Sciences University Center for Health and Healing, in Portland, reported a 61% projected reduction in water use, which includes fixture, irrigation, and cooling systems. To achieve this reduction, the site employs water-conserving fixtures, rainwater harvesting, storage and reuse of groundwater, and on-site treatment and recycling of wastewater. All wastewater is treated and reused on site for cooling, toilets, urinals, and irrigation, and a green roof reduces storm-water runoff from the site by roughly 50%. Water-conserving fixtures added negligible costs over conventional fixtures, and the initial investment of $50,000 that was required for the rainwater harvesting and recycling system was offset by a $50,000 incentive from the city. The on-site sewage-treatment system was financed by a third-party vendor and did not add up-front cost for the owner. Portland's high rates for wastewater treatment—which are designed to fully reflect the capital costs of wastewater infrastructure improvements—make additional investments in water-saving technologies cost-effective.

BENEFITS OF WATER AND WASTEWATER SAVINGS

U.S. water and wastewater rates increased an average of 4% per year between 1996 and 2006, and experts expect rate increases to accelerate over the next decade,[72] largely because of needed renovations of aging infrastructure. A 2007 survey found that in the previous five years, municipal water rates had increased 27% (or roughly 5% annually) across the United States; during the same period, rates increased 32% in the United Kingdom (U.K.) and 58% in Canada.[73] (Appendix J describes recent water-rate increases in a sampling of locations across the country.)

Three factors are expected to cause water prices to increase at rates well above inflation: the need for capital-intensive expansions and renovations of water systems; increasing populations in regions with limited water supplies; and the increasing cost of the energy used to treat and transport water. A 2002 report by the U.S. Environmental Protection Agency (EPA) estimated that nationally the gap between investment needs and planned spending was $140 billion for clean water systems.[74] As federal spending on water infrastructure has decreased, a greater share of infrastructure costs is borne by states, municipalities, and, ultimately, water customers.[75]

For our study, we assumed that water rates will rise at 3% above inflation, or 5% annually. Figure 1.17 shows the present value of 20 years of water savings for various building types. Water savings also vary by LEED level; for green schools, for example, estimated water savings range from approximately $1/sf for certified schools to $2/sf for Platinum schools. The median water-use reduction (39%) for the entire data set was used to estimate savings by building type. Benefits range from $0.50/sf for offices and religious-assembly buildings to over $1.50/sf for health care facilities.

FIGURE 1.17 Present Value of 20 Years of Water Savings by Building Type

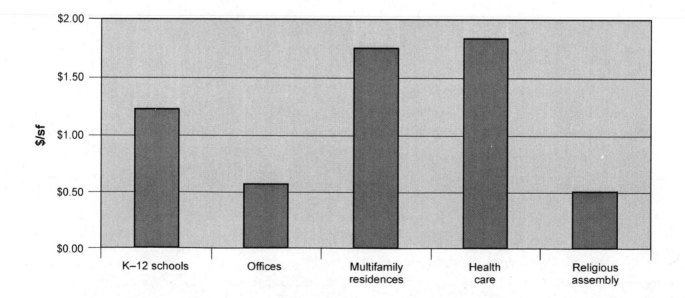

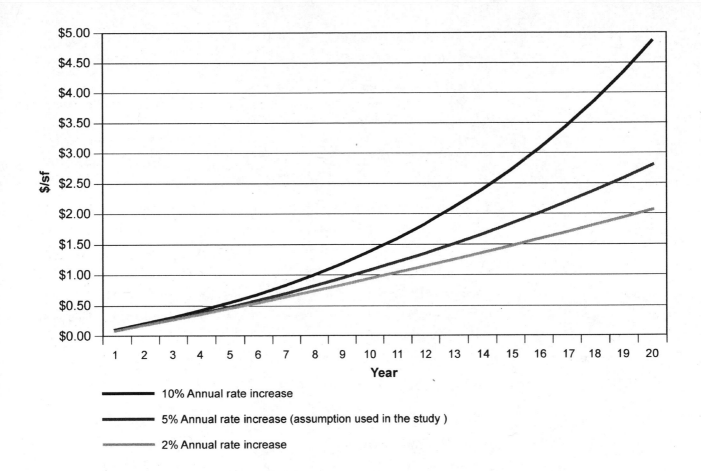

FIGURE 1.18 Cumulative Water Savings in Green Schools: Sensitivity to Escalation in Water Prices

Based on data from six buildings in Appendix J, the additional cost of achieving these savings can be less than 0.2% of total building costs, or less than $0.20/sf to $0.60/sf for most buildings in the data set. Using the conservative assumptions outlined earlier for estimating direct savings, the present value of direct water savings in green buildings outweighs the initial costs of those strategies for all building types. Figure 1.18 shows the sensitivity of cumulative benefits of water savings to future escalation in water prices, based on typical rates of water use in schools.

INDIRECT WATER SAVINGS

As with energy use, wide adoption of water conservation in green buildings would eliminate a portion of future water and wastewater infrastructure investments, including the need for new and expensive sources of potable water. Water conservation programs in buildings, even when financed publicly, are generally significantly cheaper than new sources of public water. For example, in 2006, the Water Conservation Program of the New York City Department of Environmental Protection undertook a financial analysis of New York City's rebate program for low-flow toilets, which had been undertaken in the 1990s:

> A 20-year net present value comparison of the toilet rebate program and equivalent expansions of the supply and wastewater systems found that the conservation program would provide a net savings of $196 million from deferring construction of new supply and wastewater treatment capacity by ten years. The cost of conserved water was estimated at $4.54 million per million gallons of water per day (MGD) saved, as compared to approximately $10 million per MGD for new supply and wastewater treatment sources.[76]

Water-use reductions also lead to energy savings: water and wastewater conveyance and treatment consume 4% of electricity used annually in the United States.[77] In some Western states, the need to pump drinking water over large distances and vertically greatly increases the energy used in water distribution. An analysis by the Natural Resources Defense Council (NRDC) of water systems in California found that if water conservation strategies were used to offset the next 100,000 acre-feet of needed municipal water supply, the savings in electricity would be enough to provide power to 25% of all households in San Diego.[78]

STORM WATER

Many green buildings reduce storm-water flow through the use of green roofs, porous pavement, rainwater harvesting, or bioswales. Over 40% of LEED-NC-certified buildings earn a LEED point for minimizing the impact of development on storm-water flow; on sites where imperviousness is already high, these buildings can decrease postconstruction runoff through measures such as green roofs and pervious pavements.[79] Two buildings in the data set, the Banner Bank Building and the Lewis and Clark State Office Building, were designed to completely eliminate storm-water runoff through rainwater capture and reuse or on-site infiltration. (The Banner Bank Building reported an overall green cost premium, including water, energy, indoor environmental quality, commissioning, and other green design strategies and technologies, of less than 1% of total building costs; see appendix C.) Heifer International Headquarters, also included in the data set, reported a 70% decrease in post-development runoff.

Although it is difficult to benchmark storm-water flow in green versus nongreen buildings, the high costs of pollution and storm-water infrastructure suggest the potential for large savings. A recent study in Washington, D.C., that modeled the impact of installing green roofs on proposed and existing roof space on large buildings in the downtown area found that storm-water flow would be reduced by 1.7% citywide, and that the total number of sewage overflow events would be reduced by 15%. In a city planning to spend $1.3 billion to reduce such overflows and to update its water and sewer systems, reducing stormwater flow through green roofs or other green building strategies could result in large savings.[80] A similar analysis undertaken for New York projects that the city would reduce city-wide capital costs for storm-water overflow infrastructure by 0.6% to 3.4%.[81]

Municipalities are increasingly recognizing the potential capital savings that can result from green buildings, and are instituting incentives or grants to help offset the costs of green roofs and other storm-water reduction strategies. However, some zoning rules still require developers to install storm-water infrastructure—such as culverts and drains—even when storm-water flow is captured or used on site, limiting upfront cost savings from green measures. Green developments implement strategies to capture or infiltrate storm-water on site over entire neighborhoods, resulting in substantial reductions in storm-water runoff and related water pollution when compared with conventional sprawl. In conservation developments, storm-water reduction strategies—including habitat restoration and reduced grading and land-clearing—produce net first-cost savings of about 25% of site development costs, based on a detailed analysis of ten ecological conservation developments including over 1,500 homes. (See section 2.8, "Cost Savings in Ecologically Designed Conservation Developments.")

1.6. Green Affordable Housing: Enterprise's Green Communities Initiative

STOCKTON WILLIAMS AND DANA BOURLAND

Enterprise's Green Communities initiative is the largest effort ever undertaken to mainstream sustainable principles and practices in the affordable-housing industry. As of the end of 2008, Enterprise had invested $615 million in grants, loans, and low-income housing tax credit equity to create healthier and more sustainable affordable housing. This investment is supporting more than 325 developments that comprise 14,500 affordable green homes across the United States. More than 60 Green Communities developments, with nearly 3,200 units, were complete by the end of 2008.

The Green Communities Criteria (GCC) were developed by Enterprise, the American Institute of Architects, the American Planning Association, the Center for Maximum Potential Building Solutions, Global Green USA, the National Center for Healthy Housing, the NRDC, Southface, and experts associated with the USGBC.[82]

The GCC apply to new construction and rehabilitation, and to multifamily buildings as well as single-family homes; they incorporate both smart site planning and green building features, such as energy efficiency and healthy homes practices. The GCC were specifically designed to enable affordable-housing developers to deliver significant, measurable benefits in terms of health, economics, and the environment, and to do so sustainably and cost-effectively.

Enterprise's ultimate goal is to make all affordable homes in the United States environmentally sustainable. Enterprise recognized when it first initiated the Green Communities project that its efforts alone will not transform the market. Mainstream financial institutions that provide construction and long-term financing, and government agencies that subsidize development and operations, must make fundamental changes to their policies to recognize and support the value of green affordable homes. Enterprise's market-transformation strategy includes actively making the case for sustainability to these important actors.

Improved performance in green affordable homes includes reduced energy and water use, increased durability, and improved resident health. If banks, for example, would provide lower-cost loans or more flexible underwriting to green residential developments based on their superior performance, it would have a profound impact on the affordable-housing industry.[83] In the years that Enterprise has worked closely with many banks and public agencies, it has become clear that both sectors are deeply interested in data that show improved building performance and operating-cost savings in green affordable-housing developments.

In late 2009, Enterprise completed an evaluation of the costs and performance of Green Communities developments relative to conventional affordable housing. This analysis reviews the experience and performance of more than 40 green affordable developments designed and built to generally uniform green building standards (the GCC). The report constitutes the most robust and comprehensive evaluation of the costs and benefits of green affordable housing to date.

In broad terms, the analysis shows that

- Affordable homes targeted to lower-income households can achieve a substantial improvement of health and environmental performance for a slightly higher but fully feasible development cost.
- Holistically green affordable homes can deliver significant and demonstrable health, economic, and environmental benefits for residents in low-income communities.
- It is cost-effective to green affordable homes: the financial benefits greatly exceed the additional cost of greening.

COSTS

Prior to the Enterprise evaluation, the most complete look at the costs and benefits of green affordable housing to date, completed in 2005, examined 16 green affordable developments around the United States. Total development costs ranged from 18% below to 9% above the costs for comparable conventional affordable housing. The average cost increase as a result of the green features was 2.4% (median 2.9%). Higher incremental costs in the properties analyzed were largely due to increased construction costs, rather than to increased design costs.[84]

Enterprise's experience, and that of a growing number of affordable-housing developers and owners, increasingly suggests that certain green methods and materials have lower first costs than conventional construction practices and can help compensate for any incrementally higher costs associated with other green features in the project. For example, properly sized heating, ventilating, and air-conditioning (HVAC) systems may be smaller and less expensive; advanced framing techniques may use less lumber; and recycling construction waste may reduce tipping fees. More broadly, denser development can save on infrastructure costs for localities.

Evidence in other building sectors and anecdotal experience in affordable housing suggest that green development costs decrease as developers gain education and experience, and increase their use of technical assistance.

AFFORDABLE ENERGY

A national survey conducted in 2005 documented the brutal choices that families living in poverty make when faced with unaffordable home-energy bills. The study found that during the previous five years, in order to pay their energy bills,

- 57% of nonsenior owners and 36% of nonsenior renters went without medical or dental care;
- 25% of nonseniors made a partial payment or missed a whole rent or mortgage payment;
- 20% of nonseniors went without food for at least one day.[85]

Green affordable housing provides substantial financial savings to low-income families by reducing energy costs. Results from two Green Community projects in Oregon—Royal Building and Clara Vista Town Homes—show energy savings of 37% and 73% respectively, compared with energy use in nearby conventional affordable-housing projects.

Many green affordable-housing projects also use more durable materials and equipment, reducing replacement costs and providing additional life-cycle financial benefits when compared with conventional affordable housing.[86]

HEALTHY HOMES

A growing body of research shows that poorly built, unhealthy buildings have direct and measurable adverse effects on physical and mental health, adding to the existing health burdens of the elderly, those with chronic medical conditions, and residents of low-income communities (which are disproportionately made up of minority populations). Low-income and minority communities are often situated in less environmentally healthy areas and experience greater rates of disease, limited access to health care, and other health-related disparities. Studies have shown that poor indoor environmental quality (IEQ) tends to interact with existing health problems and magnify health disparities, compounding already distressing conditions.[87]

A promising effort is under way at the High Point HOPE VI (Green Communities) development in Seattle, which provides an example of the benefits of improved IEQ in green homes. The community includes 35 "breathe-easy" homes, which were built for families with children who have asthma. In addition to typical green features, the breathe-easy homes employ a host of strategies to keep indoor air healthy, including positive-pressure ventilation to circulate stale, dirty air out of the homes; moisture-removing fans; insulated foundations to help control the interior climate; doormats to trap loose dirt; green cleaning supplies; and the use of low-VOC paints and cabinetry adhesives.

Research on the health impacts of moving into the breathe-easy homes revealed significant and measurable health results: in their old homes, asthmatic children experienced an average of 7.6 symptom-free days in a given two-week period, compared with 12.4 days in their new breathe-easy homes. In their old homes, study-group participants made an average of 60 trips a year to the emergency room; in the new breathe-easy homes, that figure declined to 21. Thus, compared with residents' previous homes, the breathe-easy homes reduced asthma triggers in the home environment. In addition, mold was eliminated, and caretaker quality of life improved.[88]

Nine million U.S. children under the age of 18 have been diagnosed with asthma.[89] Healthy IEQ can materially reduce asthma and associated health problems while providing a significant economic benefit, especially to low-income families. Improvements in IEQ—a key component of the Green Communities standard—can have considerable impacts by enabling families to reduce health care costs, reducing school absenteeism, and reducing loss of compensation from missed work.

Green neighborhood design can also deliver economic benefits to low-income families. A study of 28 metropolitan areas found that families with incomes between $20,000 and $50,000 spend an average of 29% of their incomes on transportation.[90] Locating affordable housing in dense, mixed-use, walkable areas—a key feature of green criteria—can significantly reduce these costs. The location of green affordable homes can reduce the environmental impacts of excessive driving and traffic congestion while providing easy access to jobs, goods, services, recreation, and public transportation. In contrast, poorly planned development can isolate low-income households in distressed areas and make it harder to access good schools and job opportunities. (For a more in-depth discussion of the transportation, health, and social impacts of green community development, see part 2, "Costs and Benefits of Green Community Design.")

Development design that creates a sense of community, encourages walking, and provides access to parks and mass transit also facilitates healthier transportation and recreation choices. Extensive research demonstrates that people who live in walkable neighborhoods walk more, weigh less, and are less likely to suffer from high blood pressure than those living in conventional developments (see section 2.4, "Transportation and Health Impacts of Green Community Design").[91]

TABLE 1.2 GREEN AFFORDABLE HOUSING

Project name	City	State	Green standard	Year built	Number of units
Southeast Phillips Creek	Milwaukie	Oregon	Earth Advantage	2005	14
Orchard Gardens	Missoula	Montana	LEED-certified	2005	35
Station Place Tower	Portland	Oregon	Portland Development Commission Green Standards	2004	176
Royal Building	Springfield	Oregon	Enterprise Green Communities	2007	33
Clara Vista Townhomes (Green Communities)	Portland	Oregon	Enterprise Green Communities	2006	44

Notes: Energy savings and green premium values for Clara Vista Town Homes are based on preliminary analyses conducted by Enterprise Green Communities that compare the project with nearby affordable-housing projects. Other green premium and energy savings values are based on code baselines and data source estimates.

A STRONGER ENVIRONMENT

Affordable housing generates significant CO_2 emissions. Nearly 30 million housing units (roughly one-third of all housing units in the country) are owned or rented by families earning $25,000 a year or less. Greening affordable housing can contribute to efforts to fight climate change. For example, a study found that weatherizing 12,000 homes in Ohio cut more than 50 tons of SO_2 and 24,000 tons of CO_2 per year, while reducing utility costs for low-income homeowners by an average of several hundred dollars per year.[92]

Through reductions in energy and water costs and improvements in health and environmental impacts, green affordable housing generally creates more valuable properties. Many residents of affordable housing enrolled in federal programs have the opportunity to gain ownership of rental properties, typically after a period of 15 years; other affordable-housing units are available for purchase at reduced prices. By increasing equity value in the long term, green design in affordable housing can give low-income homeowners an important economic boost.

Table 1.2 shows a sample of five green affordable-housing projects in the study data set. For a typical affordable-housing unit, annual energy savings would total $200 to $400. The average cost premiums reported by affordable-housing units in the data set are generally consistent with the 2005 Tellus Institute study of 19 green affordable-housing projects, and with those from the ongoing Green Communities evaluation effort.[93]

TABLE 1.2 GREEN AFFORDABLE HOUSING *(CONTINUED)*

Total cost ($/sf)	Green premium (%)	Annual energy savings (%)	Annual energy savings ($/household)[1]
Not available (N.a.)	3.5	15	133
170	0.0	43	380
171	3.5	N.a.	N.a.
153	—	37	327
106	−1.9	73	645

1 Annual household savings estimates assume average national annual energy expenditures for multifamily buildings of $884/household; see http://buildingsdatabook.eere.energy.gov/.

1.7. Health and Productivity Benefits of Green Buildings

Question: How do green buildings impact occupant health and productivity? What is the value of potential financial benefits from improved health and comfort in green buildings?

Evidence: Green buildings in the data set undertook a range of measures to improve IEQ but generally did not track health and productivity impacts. Less than 10% of the buildings in the data set reported undertaking occupant comfort surveys or provided anecdotes about improved health. No buildings in the data set documented direct measurable financial savings from the health or productivity benefits of building green. However, recent surveys have documented that occupants of green buildings are more comfortable with air quality than occupants of conventional buildings.

Bottom line: Current research suggests that green buildings improve indoor environments but generally does not indicate the magnitude of impacts. Further data and more widespread surveying of building occupants are needed to develop accurate estimates of the value of indoor health benefits in green buildings.

Americans spend roughly 87% of their time indoors.[94] The quality of indoor environments has a large impact on health, and on the effectiveness of countless activities performed in buildings. Asthma, colds, flu, allergies, sick-building syndrome, and mental health problems have all been linked to poor indoor environments.[95] Multiple peer-reviewed studies have linked improved indoor environments to greater productivity, including increases in the amount of work accomplished, better student performance, improved worker retention, reduced absenteeism from work or school, and reduced hospitalization times.[96] A growing number of managers, school administrators, homeowners, and building owners now view IEQ not only as a way to reduce the risks and costs of health problems , but also as a means of improving the quality of the time spent indoors and adding value to properties and businesses. Salaries and health-care costs dominate the budgets of most corporations and public institutions, so even small improvements in health or productivity can have substantial financial impacts—potentially much larger impacts than utility or operational savings from energy or water efficiency.

However, as of this writing, there is limited data on health and productivity improvements in green buildings specifically. Optional questions on health and productivity impacts that were included in our project survey went largely unanswered, reflecting a general dearth of monitoring of health and productivity. Health and productivity impacts are inherently more difficult to measure than energy and water consumption, and would be expected to vary not only with building characteristics, but also with the characteristics of the populations using the buildings and the activities performed there. In its November 2007 National Green Building Research Agenda, the USGBC described the need for a national program to "develop protocols to assess public health impacts of the built environment; and conduct assessments of impacts of IEQ on human performance, including research on mechanisms and types of tasks, in different building types."[97]

Providing superior IEQ remains a primary focus of green building design and construction. LEED 2009 devotes roughly 14% of possible points and two required measures (prerequisites) to IEQ-related strategies.[98] LEED buildings in the data set earned, on average, 62% of available points in the IEQ category; LEED-NC-certified buildings earned an average of 56% of possible IEQ points. Earning LEED IEQ credits variously requires the use of low-emitting paints and coatings; meeting and exceeding standards for ventilation rates and air-quality monitoring; limiting the spread of indoor air pollution from sources such as chemical storage, printing areas, and tobacco smoke; meeting construction-phase standards for protecting materials, to prevent later mold and contamination problems; flushing out pollutants prior to occupancy; providing access to views; improved natural light; and improved thermal control and comfort. According to Bob Thompson, chair of the LEED IEQ Technical Advisory Group and branch chief of the EPA's Indoor Environment Management Branch, the large majority of new, nongreen buildings do not pursue the indoor environmental performance measures characteristic of green buildings and would earn few points in the LEED IEQ category.[99] Green buildings generally have measurably better IEQ than nongreen buildings, with associated positive impacts on health and productivity.

Linkages between buildings and occupant health and productivity are complex, as are the building environments in which these linkages play out. Estimates of the financial benefits from health or productivity improvements in green buildings are characterized by a high degree of uncertainty, and should be viewed as indicative of a range of likely outcomes—not as a prediction of specific benefits. Further research on health and productivity in green buildings will be needed to develop estimates of benefits that are based on specific building types and linked to specific green building measures.

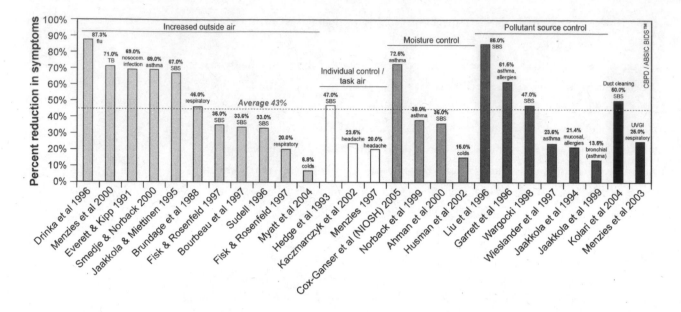

FIGURE 1.19 Health Gains from Improved Indoor Air Quality

Source: Carnegie Mellon University, Center for Building Performance and Diagnostics, 2007.

Notes: (1) For full references, please refer to www.bomaottawa.org/en/Committees/documents /Handout1BIDSDocument.pdf. (2) SBS = sick-building syndrome. (3) UVGI = Ultraviolet germicidal irradiation.

HEALTH AND PRODUCTIVITY AND INDOOR ENVIRONMENTAL QUALITY

A large number of studies have examined the effects of air quality, moisture, lighting, temperature, individual lighting and temperature control, views, acoustics, and layout on building occupants. Impacts include changes in rates and symptoms of asthma, flu, colds, and allergies; and changes in productivity, student learning, retail sales, and patient recovery. To help inform decisions about building design, the Center for Building Performance Diagnostics at Carnegie Mellon University has created a database of such studies: the Building Investment Decision Support (BIDS) tool.[100] The Indoor Air Quality Scientific Findings Resource Bank, developed by the Lawrence Berkley National Laboratory, contains summaries of the current state of research on IEQ, including impacts on health, absenteeism, and student and worker performance; discussions of implications for building design; and links to, and listings of, major papers on the topics.[101]

Figures 1.19 through 1.22 present the findings from a sampling of studies in the BIDS database that demonstrate the range of impacts on health and productivity from the types of building environments sought in green buildings. Each bar in these figures represents an independent and peer-reviewed case study or controlled experiment examining a

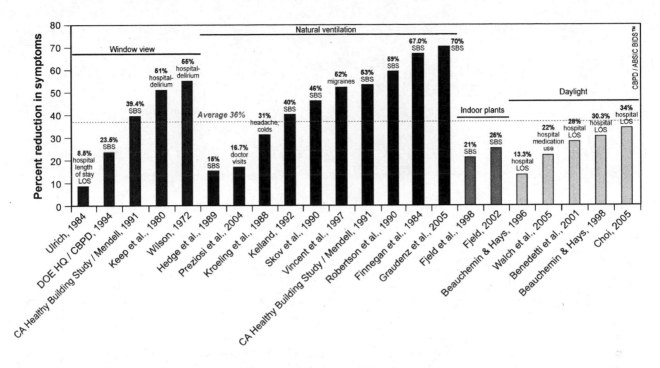

FIGURE 1.20 Health Gains from Improved Access to the Natural Environment

Notes: (1) For full references, please refer to www.bomaottawa.org/en/Committees/documents
/Handout1BIDSDocument.pdf. (2) SBS = sick-building syndrome. (3) UVGI = Ultraviolet germicidal irradiation.

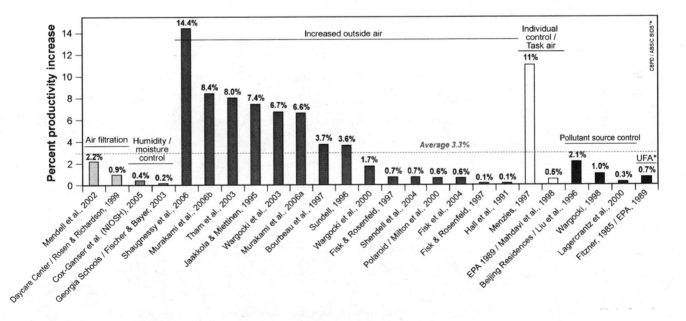

FIGURE 1.21 Productivity Gains from Improved Indoor Air Quality

Source: Carnegie Mellon University, Center for Building Performance and Diagnostics, 2007.

Notes: (1) For full references, please refer to www.bomaottawa.org/en/Committees/documents
/Handout1BIDSDocument.pdf. (2) Productivity gains are adjusted for time at task. (3) UFA = Under-floor air.

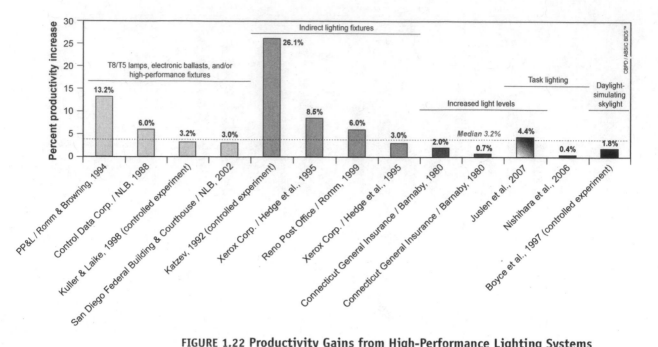

FIGURE 1.22 Productivity Gains from High-Performance Lighting Systems
Source: Carnegie Mellon University, Center for Building Performance and Diagnostics, 2007.
Notes: (1) For full references, please refer to www.bomaottawa.org/en/Committees/documents
/Handout1BIDSDocument.pdf. (2) Productivity gains are adjusted for time at task.

particular attribute of indoor environments. Health impacts include reduced symptoms from improved air quality (for figure 1.19, the average is 43% reduction in symptoms), and from access to the natural environment (for figure 1.20, the average is 36% reduction in symptoms). Each study evaluates very specific building improvements, such as increased outside air, increased control over building moisture, increased control of sources of indoor pollutants, increased access to window views, increased daylighting, and greater natural ventilation. In each study, a positive association was found between the particular attribute of the indoor environment and a specific health impact (e.g., reductions in the occurrence of asthma, headaches, colds, and sick-building syndrome).

Productivity improvements are also linked to improved indoor air quality (figure 1.21, average 3.3% improvement), improved temperature control (average 5.5% improvement), and high-performance lighting systems (for figure 1.22, the median improvement is 3.2%). These studies, it should be noted, do not specifically document impacts in green buildings. However, the design attributes addressed (such as improved ventilation, lighting, and temperature control) are common to green buildings, and are specifically required and/or rewarded by certification systems such as LEED.

Improved indoor environmental quality (IEQ) can increase productivity by reducing health problems and the associated loss of time and energy. Additional productivity improvements may result from improved worker performance attributable to greater comfort, greater control of temperature, and improved lighting and air quality.

Increased daylighting, views, and contact with nature have been linked to positive health and productivity impacts in a number of different ways for different building types. In a series of studies on the impact of daylighting, it was found that pleasant outdoor views in classrooms are strongly linked to improved student performance.[102] Daylighting has also been linked to enhanced student performance, provided that classrooms are designed to avoid the negative impacts of glare. Similarly, outdoor views have been shown to be associated with faster patient recovery in hospitals—potentially a large cost savings to green health care facilities, patients, and insurance companies.[103] Hospital stays cost, on average, $1,200 per day—meaning that a small reduction in the length of hospital stays can substantially reduce health care costs.[104] Other studies have documented an association between the presence of skylights and increases in retail sales.[105]

Many see an inherent benefit in demonstrating concern and care for green building occupants. Such impacts are difficult to trace to specific building attributes but may be significant. As Dave Wood, a science teacher at the LEED Platinum Sidwell Friends Middle School in Washington, D.C., has noted, "Building a green school sends an unequivocal message to students" about the value the school places on both the environment and the students themselves.[106] Michael Saxenian, assistant head of school and chief executive officer, observes that "by modeling a different approach to nature, the green buildings infuse our students with a sense of hope and possibility."

HEALTH AND PRODUCTIVITY IN GREEN BUILDINGS

Green building standards for IEQ have been based, in large part, on research—some of which was highlighted earlier—linking specific attributes of building environments to human impacts. How do occupant health and occupant productivity in new green and nongreen buildings compare?

Scattered anecdotal responses to the project survey indicate generally positive impacts, but few of the respondents had systematically documented increases. For example, officials at Third Creek Elementary, a LEED Gold school in Statesville, North Carolina, reported that in the two years after the move to the green school, 80% of the school's 700 students tested at grade level for reading and math, up from 60% in the three years prior to the move. Officials attribute this improvement to the green school.[107]

Surveys of occupant comfort and satisfaction provide one means of assessing the success of IEQ strategies in green buildings. The Center for the Built Environment (CBE) at the University of California, Berkeley, implemented a survey of occupant satisfaction with the built environment in a range of buildings, both green and nongreen. Analysis of

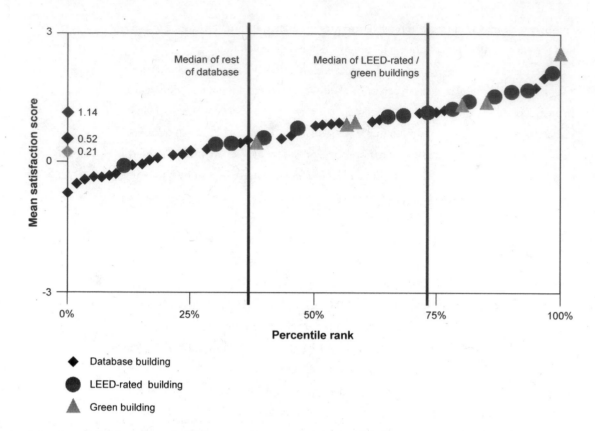

FIGURE 1.23 Satisfaction with Air Quality in Green versus Conventional Buildings

Sources: S. Abbaszadeh et al., "Occupant Satisfaction with Indoor Environmental Quality in Green Buildings," *Proceedings of Healthy Buildings,* vol. 3 (2006): 365–370; available at http://repositories.cdlib.org/cedr/cbe/ieq/Abbaszadeh2006_HB/.

Note: Comparison is between green buildings (n = 21) and other buildings in the Center for the Built Environment database (n = 35); all buildings were younger than 15 years. Mean satisfaction score with air quality for LEED-rated/green buildings is 1.14 vs. 0.52 for non-green buildings younger than 1.5 years and 0.21 for all ages of non-green building in CBE database (see diamonds on left axis).

responses from 21 green and 160 nongreen buildings found that green building occupants were on average twice as satisfied with air quality, thermal comfort, and the overall building (median score of 1.14 for green buildings vs. 0.52 for non-green buildings), but revealed no statistical difference in satisfaction with lighting and acoustics between green and nongreen buildings. When the comparison group of nongreen buildings was restricted to only those built in the prior 15 years, there was a statistically significant increase in satisfaction with air quality in green buildings, though no statistically significant differences for other IEQ categories (see figure 1.23).[108]

Detailed survey results in the lighting and acoustics categories suggest that lighting controls in green buildings did not significantly increase occupants' ability to control lighting levels. Open layouts in LEED offices, which are often created to increase access to views and daylighting, sometimes led to acoustic problems.[109]

Gary Jay Saulson, Director of Corporate Real Estate, PNC Financial Services Group

PNC Financial Services Group, now the fifth-largest U.S. bank by deposits, has been involved in the construction of green buildings since the late 1990s.[1] During this time, PNC has maintained a focus on measuring results and understanding the cost, energy, and productivity impacts of building green. For our first green building, the 647,000-square-foot First-side Center, in Pittsburgh, which was begun in 1998 and opened in 2000, PNC meticulously tracked every measure of productivity and saw improvements across the board.

By the time we had completed our first large green building, one of our companies, PNC Bank, was in the midst of an aggressive expansion program that involved building new branches. Taking advantage of this opportunity, we decided to develop a new green model for building PNC branch banks. We wanted to design a branch that addressed PNC's commitment to our employees, the communities in which we do business, our customers, and our shareholders. We wanted to design an iconic building that was green, and in which people could thrive. The process began with an eco-charrette in which we talked about dreams and aspirations and explored building systems, energy consumption, lighting schemes, and design. We developed a Green Branch prototype that could be replicated in a variety of climates and locations, and that would provide high-quality customer service while delivering high-performance building results.

As of January 2009, PNC had built 53 LEED-certified Green Branches. In order to validate the success of our Green Branch initiative, we engaged Paladino and Company to conduct a post-occupancy evaluation of ten Green Branches and ten legacy (existing nongreen) branches. We faced a number of questions: Has the Green Branch improved performance? Has it improved predictability? Has it improved occupant satisfaction? To answer these questions, we compared utility data from legacy and Green Branch buildings, pairing legacy and Green Branches in similar climates and normalizing the data by building size to remove the impacts of climate and scale on energy use. A survey of occupant satisfaction was also conducted using a Web-based post-occupancy evaluation protocol developed and administered by the Center for the Built Environment (CBE) at the University of California, Berkeley.[2]

The Green Branch was found to reduce both overall energy use and the variability of energy use, lowering maintenance and energy costs across our portfolio of branch banks. By reducing exposure to energy cost increases, Green Branch buildings realized direct energy savings of more than 34% when compared with legacy branch buildings. Standard deviation of energy use in Green Branch buildings was reduced by about 85% (see figure 1), a tremendous improvement in standardizing the achievement of building-performance goals. In one typical Green Branch building, actual water use was more than 44% lower than code and 29% lower than the level of use in a comparable legacy branch building.

Green Branch occupants reported significantly higher satisfaction with the office layout, air quality, lighting, temperature, and acoustics than occupants of legacy branches; reported satisfaction was also significantly higher than the average results in the CBE's extensive database of occupant satisfaction surveys (see figure 2). Green Branch occupants reported greater satisfaction with the building than 98% of respondents in the CBE's database, which included primarily nongreen office buildings. These results provided striking

(continued on page 54)

FIGURE 1 Energy Use in Green Branch and Legacy Branch Buildings
Note: Standard deviation was reduced by 85%.

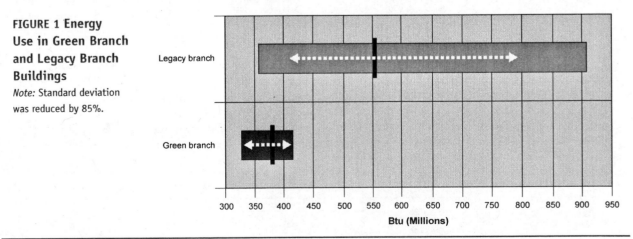

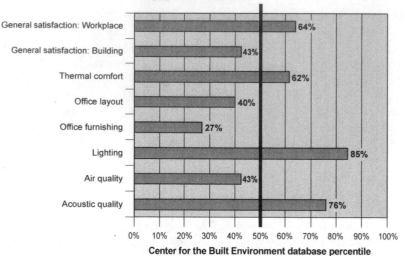

Legacy median response

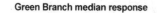

FIGURE 2 Occupant Satisfaction Scores: Green Branches and Legacy Branches

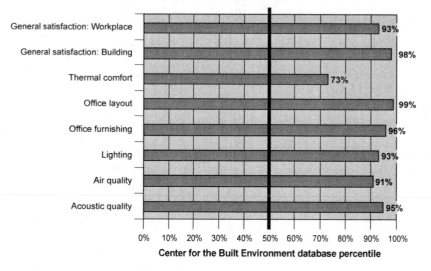

Green Branch median response

carefully considering every building component allowed us to find opportunities for cost savings. For example, hot water in legacy branches was provided by standard, 60- to 80-gallon residential hot-water heaters. Banks need much less hot water than households, however, so in our new branches we use six-gallon heaters, reducing both costs and energy use. By purchasing plywood directly from certified lumber companies, we were able to get Forest Stewardship Council–certified wood at a lower price than conventional plywood for our Green Branches. We also volume-purchased wall panels, glazing, and steel. PNC Green Branch banks are built for roughly $100,000 less than branches built by one of our competitors. We maintain a list of technologies that we would like to use in future buildings but that may not yet be cost-effective, including photovoltaics, geothermal heating, and reclamation and treatment of rainwater and wastewater.

Most importantly, from a business perspective, green has now become a central part of the PNC brand. Our green buildings have brought us a tremendous amount of community

evidence of PNC's success in delivering higher-performance buildings that positively impact people, the planet, and the bottom line. A next step in our process should be to analyze employee absences and turnover at Green Branches, and to compare the results to data derived from our legacy branches.

PNC's Green Branches were built at the same initial cost as the legacy branches. In other words, these benefits have been achieved without any cost premium. In some areas,

recognition and positive media attention. What began as an effort to green one building has become an ethic and a practice embraced by the entire company, from the chief executive officer to our bank tellers, impacting employees and customers alike. PNC employees have a greater sense of pride in the buildings they work in, and a great point of connection and conversation with the customers who walk into a Green Branch.

1. See www.pnc.com (includes PNC Bank).
2. See www.cbe.berkeley.edu.

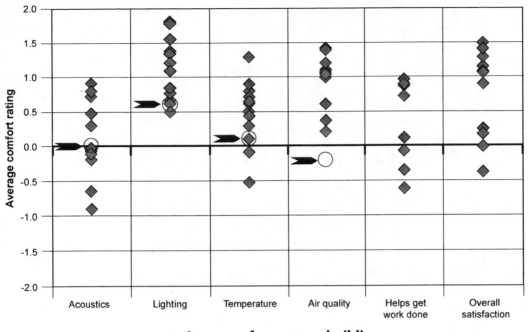

Average of non-green buildings

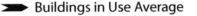 Buildings in Use Average

+2 = most comfortable
-2 = most uncomfortable

FIGURE 1.24 Occupant Comfort in LEED Buildings
Source: New Buildings Institute, "Energy Performance of LEED for New Construction Buildings: Final Report," March 4, 2008 (www.newbuildings.org/downloads /Energy_Performance_of_LEED- NC_Buildings-Final_3-4-08b.pdf). *Note:* Buildings in Use Average represents average occupant comfort rating in a large database of nongreen buildings.

As of January 2009, PNC bank had constructed 53 LEED-certified branch buildings. CBE's occupant-satisfaction survey, conducted in five of PNC's Green Branches, found that occupants' overall satisfaction with the building and with lighting, temperature, acoustics, and air quality was significantly higher than that of occupants of PNC's non-green branches. Compared with CBE's database of survey results (primarily from non-green buildings), the results of occupant surveys in PNC's Green Branches were above the 90th percentile for seven out of eight survey categories, showing improved comfort among Green Branch occupants (see "Perspective: Birth of the Green Branch Bank").[110]

The Work Environments Research Group at the University of Montreal implemented a similar occupant survey in a large number of office buildings, and compiled responses from over 1,000 cases in its Buildings in Use (BIU) database.[111] A New Buildings Institute study on post-occupancy performance for LEED buildings implemented a version of the University of Montreal survey in 12 recent LEED buildings, including four buildings also included in the study data set: the Colorado Department of Labor and Employment, The Henry, Hermitage Elementary, and the Lewis and Clark State Office Building. Figure 1.24 compares the average scores in primary IEQ categories for these 12 buildings to the scores from thousands of responses from office buildings in the BIU database. Median scores from the 12 buildings show higher-than-average occupant comfort in LEED buildings.[112] It should be noted, however, that because of the small sample size, the NBI did not perform statistical analysis on the results.

Clearly, there is a need for more green buildings to survey occupant comfort, both to better document performance in achieving IEQ goals and to refine high-performance design approaches and green building standards. While LEED-NC does not include any credits related to acoustics, LEED for Schools, released in spring 2007, includes two potential points for improved acoustics—a response to early indications that acoustics appears to be an area in which the first generation of green buildings does not show improved performance compared with conventional design.

ESTIMATING HEALTH AND PRODUCTIVITY BENEFITS

Occupant surveys described in the two previous sections provide evidence that greening a building improves indoor environments, but say little about the magnitude of health and productivity improvements in green buildings. A 2006 review of the evidence on health and student performance by the National Research Council, for instance, found a general lack of "well-designed, evidence-based studies concerning the overall effects of green schools on human health, learning, or productivity or any evidence-based studies that analyze whether green schools are actually different from conventional schools in regard to these outcomes."[113] Even if research establishes the magnitude of health gains in green buildings (e.g., reduced symptoms, disease prevalence, or absenteeism), the value of these improvements to owners and occupants depends on the particular building type and on maintenance and occupancy characteristics. Past efforts to quantify the value of health and productivity improvements in green versus conventional buildings have found a range of impacts, from less than $10/sf to greater than $50/sf over 20 years.[114]

So how can building owners and occupants assess the value of pursuing healthy indoor environments in green buildings for the purposes of informing current design decisions? One approach would involve *assuming* modest health improvements and estimating the monetary value of such improvements. For instance, if annual health costs were estimated at $3,000 per occupant and could be reduced by 1% through improved IEQ, this improvement would be valued at $30 per employee in the first year.[115] (As a point of reference, BIDS studies on health gains from improved IEQ, represented in figure 1.19, showed an average of 43% improvement in symptoms.)

Table 1.3 shows the range of estimated present value for the health and productivity impacts described in this and the two previous sections. In a building with 230 square feet per employee, the present value of a 1% reduction in health costs over 20 years would be

TABLE 1.3 ESTIMATED HEALTH BENEFITS OF GREEN BUILDINGS: ILLUSTRATIVE EXAMPLES

Potential impact	Estimate method	Present value of 20 years of impact ($/sf)
Reduced health care costs for green building occupants and employers	Assumed 1% reduction of $3,000 per occupant in health care costs	$2
Reduced respiratory infections, allergies, and asthma	Scaled-down national savings estimates[1]	$1
Productivity increases from improved work environment	Assumed 0.5% improvement, based on $50,000 in annual employee value	$13
Productivity increases from reduced sick-building syndrome	Scaled-down national savings estimates[1]	$12–$35

1 Based on W. J. Fisk, "Health and Productivity Gains from Better Indoor Environments and Their Implications for the U.S. Department of Energy," 1999 (available at www.rand.org/scitech/stpi/Evision/Supplement/fisk.pdf).

approximately $2/sf, assuming a 7% discount rate and a 3% annual increase in health care costs. The savings would accrue to both occupants and employers. Thus, for an employee who occupies roughly 230 square feet of office space and earns an annual salary of $50,000, if productivity can be improved by 0.5% through reduced absenteeism or improved work results in green buildings, this benefit would be worth roughly $13/sf over 20 years, using a 7% discount rate and a 3% annual salary increase (see the third row of table 1.3). (BIDS studies represented in figure 1.21 show a median productivity increase of 3.3% from improved indoor air quality alone.)

Alternatively, estimates of productivity benefits for green buildings could be based on improved revenues or sales attributable to IEQ improvements. Yet another method of valuing potential health benefits in green buildings would involve scaling national estimates of the direct and indirect costs of poor indoor air quality (e.g., respiratory illness, allergies, and asthma) to the level of per-square-foot impacts, using an assumption for average building space per person. A 1999 national review conducted by William Fisk, of the LBNL, estimated that improving indoor air quality in buildings across the country could produce $7 to $18 billion in savings from reduced respiratory illness, allergies, and asthma—taking into account both direct and indirect savings.[116] Adjusting these estimates for inflation to 2007 and assuming 1,000 square feet of combined residential and office-building space per person suggests that 20 years of health savings from improved indoor air quality would be valued at up to $1/sf for reduced respiratory illness, allergies, and asthma (see the second row of table 1.3).

The Fisk analysis also estimated that reduced sick-building syndrome would yield $10 to $30 billion in potential productivity benefits nationally, or $150 to $460 for each of the roughly 65 million U.S. office workers. Assuming this benefit could be achieved by providing 230 square feet of green office space per office worker in the United States, the present value of savings over 20 years would be roughly $12 to $35/sf (see the fourth row of table 1.3). The savings evaluated by Fisk were based on a set of IEQ measures similar to those implemented in green buildings, so these estimates suggest a range of potential health benefits in green buildings, including avoided direct medical costs and avoided indirect costs (e.g., lost work time and missed days of school).

Additional surveys of health symptoms or productivity metrics related to specific measures in green and nongreen buildings will be required to achieve more evidence based, accurate estimates of improvements in green buildings. The benefits estimates provided here are intended to be indicative and are not precise; further research will be needed to better quantify health benefits in green buildings.

1.8. Green Health Care: Assessing Costs and Benefits

ADELE HOUGHTON, ROBIN GUENTHER, AND GAIL VITTORI

Calculating the first costs and associated benefits of a green health care facility is a challenging and important undertaking.[117] Unlike the office-building sector, the health care sector has priorities—human health, and patient, workplace, and environmental safety—that are nested within a broad organizational mission focused on healing and stewardship. This important distinction creates significant challenges in defining first-cost premiums and benefits. First, this values-laden context blurs the distinctions between design elements that represent added cost to the construction budget and those that are increasingly acknowledged as standard practice for the health care sector. Second, efforts to conduct a statistically significant analysis of green premiums are hampered by the fact that the pool of completed green health care facilities (particularly green hospitals) is still relatively limited, albeit growing rapidly.

This section assesses first-cost premiums in the health care sector; it is based on data submitted by, and interviews with, 13 LEED-certified and LEED-registered health care project teams in the United States. The projects ranged in size from 28,000 to 470,000 square feet, and were completed between 2003 and 2009. Because of differences in size, energy intensity, and program complexity among the sample projects, this research dis-

FIGURE 1.25 First-Cost Premiums for Health Care Facilities in the Study Project, after Incentives, by Actual or Anticipated LEED Certification Level
Note: Denver Health Pavilion for Women and Children did not track first-cost premiums and is therefore not included in this chart.

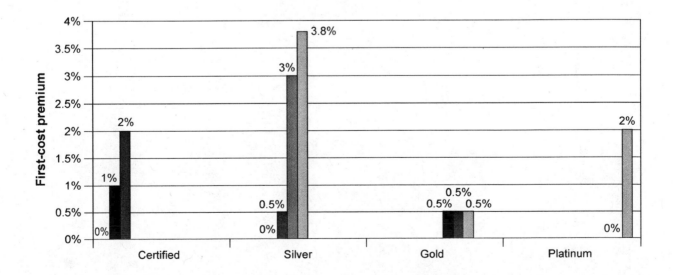

tinguishes between three medical building types: acute care hospitals, outpatient buildings, and mixed-occupancy facilities.

The subject projects reveal the complexity of estimating first-cost premiums within this sector. Overall findings include the following:

- First-cost premiums range from 0% to 5% before financial incentives are accounted for, and from 0% to 3.8% after financial incentives are included (see figure 1.25). (Incentives included philanthropic gifts, grant programs, and public or utility incentive programs.)
- First-cost premiums do not directly correlate with LEED certification level (see figure 1.25). Indeed, in keeping with the findings of some other studies, health care facilities that achieve LEED Gold or Platinum certification do not bear higher first-cost premiums than those at the LEED-certified or Silver levels.
- Projects that achieved LEED certification in the early 2000s indicated higher premiums than those that achieved certification later. This finding is consistent with other studies: in general, first-cost premiums are decreasing over time (see figure 1.26).

A closer analysis reveals a range of factors that contribute to the variability of reported first-cost premiums:

FIGURE 1.26 First-Cost Premiums for Health Care Facilities in the Study Project, after Incentives, by Year

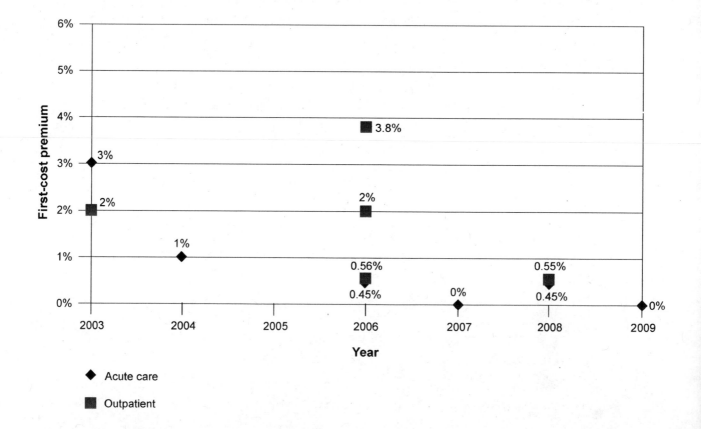

- Components of the premium calculation
- Methodologies used to calculate the premium cost
- Total amount of grants or incentives (including utility rebates) awarded to projects. (All projects that received financial incentives used those incentives to reduce the first-cost premium.)

BACKGROUND

The health care sector spans a range of building types, including hospitals, outpatient clinics, medical office buildings, and labs.[118] Licensed acute care hospitals (accounting for only 6% of the total number of health care facilities, but 60% of total square feet) are particularly complex.[119] According to the 2003 CBECS, inpatient hospitals, with an average square footage of about 241,000 square feet, are approximately 16 times larger than a typical office building; they also have over twice the energy intensity per square foot (188,000 Btu/sf versus 93,000 Btu/sf).[120] The energy intensity of outpatient facilities, such as medical office buildings, is similar to that of office buildings (95,000 Btu/sf), but the average size is slightly smaller (10,000 square feet). Given the range of energy intensity within the health care sector, it is essential, when comparing construction costs and energy intensity across health care projects, to identify the specific types of health care activities incorporated in the building program.

What drives the energy intensity of acute care hospitals? Acute care facilities are complex, stringently regulated buildings that operate 24 hours a day, seven days a week, and depend on many tiers of mechanical and electrical health and safety systems. In addition to being subject to strict regulations covering mechanical ventilation and air-change rates, hospitals contain extensive diagnostic equipment operating around the clock and generating sizable cooling loads. Thus, efforts to maximize efficiency through strategies such as "right-sizing" equipment must incorporate redundancies to safeguard critical services and to allow for future incremental growth and changes in technology.

DEFINING GREEN HEALTH CARE PROJECTS

Currently, two tools are available to green health care construction projects: the LEED rating system and the Green Guide for Health Care, a voluntary, self-certifying tool kit for best practices in green design, construction, and operations. As of October 2008, more than 160 health care projects had registered with the Green Guide; some of these overlap with the 340 health care projects registered with LEED. To ensure that the projects included in this survey had been documented as green by a third party, all the selected projects are LEED-registered projects.

As of October 2008, 35 health care projects had been certified through LEED. Of these, eight project teams were interviewed on the subject of green building premiums (specialty hospitals, long-term care facilities, and medical office buildings with less than 10,000 square feet were excluded). In addition to these eight certified projects, the data set includes five projects awaiting certification, for which construction is complete and costs have been summarized—enabling the study to obtain information on five additional acute care hospitals and to offer data on recently completed projects. The new construction projects range from large to small and include eight acute care hospitals, two mixed-occupancy buildings, and three ambulatory care facilities.

FIRST-COST GREEN PREMIUMS

The projects varied significantly in first-cost premiums and in the components that made up the premium. Eight projects received utility incentives or dedicated funding that was applied to reduce projected premiums.

Actual building components included in the premium varied widely, but aggregate premiums fell within a fairly narrow range. Projects that reported no cost premium described a design process in which the building budget was established and the team successfully integrated sustainable strategies within the overall construction budget, thereby eliminating the need to track premiums and savings on a line-item basis. Three of the 13 projects reported zero premiums, based on having delivered the project within the established budget; one project, the Denver Health Pavilion for Women and Children, did not track or report first-cost premium data. The availability of grants and incentives influenced the choice of sustainability strategies and the level of achievement. In the subject projects, the availability of financial incentives permitted increased expenditures.

The 13 projects demonstrate significant variability in the set of design elements identified as contributing to first-cost premiums. The elements isolated for inclusion were derived from the following sources:

- Strategies linked to financial incentive programs, ranging from energy-efficiency measures that qualified for utility rebates to green roofs and bioswale installations that garnered environmental grants. Some of these strategies would have been incorporated into the project with or without grant funding, but the existence of the incentive places them in the category of first-cost premium.
- Components that went above and beyond what the owner considered baseline green building practice. The wide variability of these features supports the idea that "baseline" is both a subjective and a moving definition. These items included green material substitutions that carry first-cost premiums, such as floor-

ing, cabinet substrates, and certified wood; enhanced filtration systems; and ventilation systems that rely on 100% outside air.

• Strategies that deliver operational savings but carry first-cost premiums, ranging from energy-efficiency measures to enhanced control systems and enhanced building-envelope components.

Project teams had varying opinions about which elements should be included in the green premium. For example, the earliest projects reported green premiums associated with low-VOC materials, while today such products are incorporated as standard practice.

OCCUPANT SATISFACTION AND COMFORT

Projects that gave priority to operational efficiency were more likely to incorporate strategies with direct financial payback, such as energy- and water-efficiency measures. Energy efficiency is generally the first place most health care facilities look for a concrete return on investment from green building strategies. Compared with the ASHRAE 90.1 baseline, acute care hospitals in this survey reported projected energy savings ranging from 0% to 50%; among outpatient and mixed-occupancy facilities, savings ranged from 15% to 58%. Because many jurisdictions exempt health care facilities from meeting local energy codes that are based on ASHRAE 90.1, these facilities may well outperform other

FIGURE 1.27 Modeled Energy Savings for Health Care Facilities in the Study Project, by Building Size

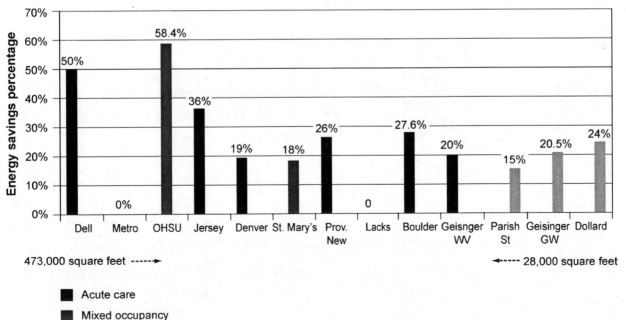

473,000 square feet ------▶   ◀----- 28,000 square feet

■ Acute care
■ Mixed occupancy
▨ Outpatient

buildings in their sector by an even higher margin than these projections suggest (see figure 1.27). (For a discussion of the value of energy savings in green buildings generally, see section 1.3, "Energy-Use Reductions.")

In general, the subject projects' modeled reductions in energy demand were 30% below code. For the most part, the projects achieved reductions by optimizing the design of conventional mechanical systems. Dell Children's Medical Center of Central Texas, in Austin, and the OHSU Center for Health and Healing, in Portland, Oregon, incorporated combined cooling, heating, and power systems, displacement ventilation, and other emerging technologies to reach greater reductions in energy use—50% and 58.4%, respectively.

Nine of the 13 projects in the data set achieved reductions of 20% or more in the use of potable water in fixtures. The Oregon Health & Science University Center for Health and Healing instituted a comprehensive water-use reduction strategy to achieve over a 60% reduction in potable water use in fixtures. While reductions in fixture water use are laudable, the resulting water savings may not be as significant in acute care hospitals, where roughly 60% of water use is for building processes such as cooling sterilizing equipment and operating the mechanical systems.

Currently, the most cost-effective large-scale water-efficiency strategies involve recycling process water through a closed-loop system and reusing process water for limited irrigation and to service the cooling tower. All of the surveyed projects reduced potable water use for landscape irrigation by at least 50%, with 100% reduction in all the outpatient projects. (For a discussion of the value of water savings in green buildings generally, see section 1.5, "Water-Related Savings.")

RELATED OPERATIONAL BENEFITS

While post-occupancy evaluations of the benefits of green health care facilities continue to be sparse, several projects in this survey have begun tracking the accuracy of projected operational benefits. Geisinger Health System identified the operational savings associated with green buildings as the primary motivation for pursuing LEED certification. In 2008, two years after opening its doors, Providence Newberg, a LEED Gold project, hired the Center for the Built Environment at the University of California, Berkeley, to conduct a third-party post-occupancy evaluation. As Richard Beam, director of energy management services in the Office of Supply Chain Management at Providence Health and Services, explains,

> The report's findings provide hard evidence that the Newberg LEED Gold Building is the new standard for health care operations. . . . It's telling us that all these things we hoped would happen—that we would have higher employee satisfaction, higher patient satisfaction, better retention of clinical employees and nurses, better physician recruitment—have scored higher due to the nature of this building.

THE WORK AHEAD

The question of green building premiums in health care is complex and challenging, and deserves continued study and tracking. Many of the health care organizations in this survey did not pursue green building for financial reasons; instead they were motivated by values of leadership and community benefit. Similarly, a larger moral purpose is often the primary motivation when religious organizations opt to build green (see part 3, "Communities of Faith Building Green"). However, one health-care-specific lesson arose in almost every project: the more emphasis the owner placed on the connection between the benefits of building green and occupant health and safety, the more sustainable building elements were integrated into the project as base building features.

Green building first-cost premiums appear to be lower than many in the industry perceive them to be, and appear to be decreasing over time. Green elements previously viewed as better-than-standard are being included in projects as a new baseline. The work ahead is to develop a more comprehensive model for tracking costs and savings from green design. As this model emerges, it will need to reflect the health care sector's triple-bottom line values, and its commitment to occupant health and safety, operational efficiency, and community benefit.

1.9. Employment Benefits of Green Buildings

Question: How many jobs do green buildings create when compared with conventional buildings? What is the value of employment impacts resulting from green design?

Evidence: A range of state and national analyses show increased job creation from investments in energy efficiency, renewable energy, and recycling.

Bottom line: Green building creates significantly more temporary and permanent employment than conventional design and construction.

Widespread adoption of green, high-performance building design would result in a large increase in employment creation when compared with conventional construction. Energy efficiency, renewable energy, and construction-waste recycling are three components of green building practice that are significantly more labor-intensive than conventional building construction and operation. A review of existing state and regional energy-efficiency programs and a macroeconomic model used to compare the impacts of green buildings to those of conventional, inefficient buildings indicate that over 20 years, green building yields an employment benefit of roughly $1/sf. However, this figure reflects only part of the employment impact of green buildings, and therefore underestimates the actual employment benefits of greening.

ENERGY EFFICIENCY AND ON-SITE RENEWABLE ENERGY

Green buildings create more jobs than conventional, inefficient buildings, in part because they shift resources away from fossil-fuel consumption and toward efficiency improvements such as increased insulation, higher-performance windows, and efficient building systems. By contrast, conventional buildings increase dependence on capital-intensive, low-labor energy industries that may rely on imported fuel, some of which comes from unstable and undemocratic countries. The construction, installation, and maintenance of energy-efficiency measures and renewable-energy systems in green buildings creates relatively skilled and professional jobs, most of which are not subject to outsourcing, providing direct economic benefits to state and local economies.

Policy makers at all levels have a vested interest in understanding the broader economic impacts of different kinds of public and private investment, including green buildings and technologies. The 2009 federal stimulus bill (the American Recovery and Reinvestment Act) was geared, in large part, toward the creation of U.S. jobs, and includes a significant focus on green jobs. Funding streams include $4.5 billion for energy retrofits of federal buildings, $6.3 billion in energy conservation grants, and $500 million directly to green jobs training and workforce development.[121]

A 2008 report by the ACEEE estimated that energy efficiency in the building industry already supports more than 1 million U.S. jobs, including 332,000 in commercial construction and renovation.[122] A 2008 report by the Center on Wisconsin Strategy noted that

> substantially reducing waste of energy through systematic retrofitting and upgrading of residential and commercial buildings is a key area where environmental and equity agendas can come together to create good jobs. The work requires a multi-skilled, local workforce that cannot be outsourced, and it feeds a building-materials industry that is still largely domestic.[123]

A number of recent studies have explored job-creation impact from large-scale shifts to energy efficiency and renewable-energy technologies, including green, high-performance buildings.

- A 2007 report by the ACEEE examined the economic impact of meeting Texas's growing energy demand with energy efficiency and renewable-energy strategies, requiring a cumulative $50 billion investment statewide between 2008 and 2036. This investment was found to produce small losses in profits for traditional utilities by 2036[124]—and, over the same period, to result in a net gain of 38,300 job-years statewide, and a $1.7 billion net increase in wages.[125]
- A 2004 analysis by Black & Veatch examined the impact of meeting a renewable portfolio standard (which requires a certain percentage of electricity to come from renewable energy) in Pennsylvania and found that 85,000 net job-years would be created relative to a business-as-usual scenario.[126]
- A 2000 report by the U.K. Association for the Conservation of Energy analyzed the impact of energy-efficiency initiatives, including improvements in building codes and programs that foster the use of efficient HVAC systems, and found an average direct net employment impact of 8 to 14 job-years created for every $1 million of investment in energy efficiency. Jobs created ranged from unskilled labor to skilled trades, engineering, and management.[127]

Table 1.4 shows projected job creation in relation to investment in energy efficiency in a sampling of recent analyses. For every $1 million shifted from conventional energy generation to energy efficiency and renewable investment, all the studies show a net increase in job creation, ranging from 1 to over 50 job-years.

Study	Focus of analysis	Investment (millions)	Projected net job-years created[1]	Projected net job-years created per million $ invested
Black & Veatch, 2004[2]	Meeting renewable portfolio standard in Pennsylvania	$1,230	85,000	69
ACEEE Midwest study[3]	Investing in natural gas and electricity efficiency	$1,100	66,260	60
Center for American Progress[4]	Efficiency, renewable energy, biofuels, and public transit	$100,000	2,000,000	20
Apollo Jobs Report[5]	High-performance building tax credits, financing, research and development, updated building codes	$89,900	827,260	9
ACEEE Texas study[6]	Meeting Texas's energy needs with renewables and efficiency	$50,000	38,300	1

1 One job-year equals one year of full-time employment for one person.

2 Black and Veatch, "Economic Impact of Renewable Energy in Pennsylvania," Heinz Endowments, Community Foundation for the Alleghenies, 2004 (www.cleanenergystates.org/library/pa/PA%20RPS%20Final%20Report.pdf).

3 Kushler, Martin, Dan York, and Patti Witte, "Examining the Potential for Energy Efficiency To Help Address the Natural Gas Crisis in the Midwest," American Council for an Energy Efficient Economy Report Number U051, January 2005 (http://aceee.org/pubs/u051full.pdf?CFID=3835657&CFTOKEN=98453078).

4 John Podesta, "Green Recovery: A New Program to Create Good Jobs and Start Building a Low-Carbon Economy," Center for American Progress, 2008 (www.americanprogress.org/issues/2008/09/green_recovery.html).

5 Apollo Alliance, "New Energy for America: The Apollo Jobs Report," 2004 (www.apolloalliance.org/downloads/resources_ApolloReport_022404_122748.pdf).

6 John (Skip) Laitner, R. Neal Elliot, and Maggie Eldridge, "The Economic Benefits of an Energy Efficiency and Onsite Renewables Strategy to Meet Growing Electricity Needs in Texas," ACEEE, 2007 (www.aceee.org/pubs/e076.htm).

In essence, the employment and economic impacts of green building derive from two sources: additional initial investment in energy efficiency (and other green technologies), and energy savings throughout the life of a building; both sources free up money to be spent in more productive, labor-intensive parts of the economy.

MODELING THE EMPLOYMENT IMPACTS OF GREEN BUILDING
Subsection by SKIP LAITNER

An input-output model is a set of economic accounts that illustrates how consumers and businesses buy and sell to each other, illuminating the ways in which changes in spending can affect the economy both directly and indirectly. For example, input-output models have shown that for every $1 million spent on utility bills, energy-related services support three to four job-years, directly and indirectly. Spending that same $1 million on green building supports roughly 8 to 12 job-years, directly and indirectly. Hence, a shift to energy efficiency would create a significant net employment benefit for the economy.

Table 1.5 illustrates how the economy might benefit from a $1 million investment in green building: it shows the 20-year change in spending, multiplied by the appropriate job multiplier (i.e., labor intensity) of each sector. Investing in the construction of green buildings means that

- More money will be spent in construction.
- Less money will be spent buying energy from utilities.
- In the short term, consumer spending will decrease because of increased spending on construction; in the medium and long term, consumer spending will increase, as money is freed up by energy savings.

The table also shows that the $1 million investment would generate a *net* gain of 16.4 job-years from the increased investment and the more productive use of energy and water. That translates into an average annual increase of about 0.8 jobs each year for 20 years.

To estimate the magnitude of employment impacts in green buildings, we ran a dynamic input-output model, based on a typical green building in the study data set: 92,500 square feet, with a green premium of $3.50/sf.[128] We assumed that the combined annual savings on utility and water bills are on the order of $0.50/sf, a figure that is consistent with findings in this book. We further assumed that the building occupant borrows money

TABLE 1.5 TWENTY-YEAR NET ECONOMIC IMPACT OF A $1 MILLION INVESTMENT IN —— GREEN BUILDING IMPROVEMENTS: ILLUSTRATIVE EXAMPLES

Spending category	Impact	Amount (millions)	Job multiplier[1]	Job impact (job-years)
Construction	Green premium increases construction spending	$1.0	12	12.00
Consumer spending	Because of the green premium, consumers spend less in the short term	$−0.6	11	−6.60
Consumer savings	Because of energy savings, consumers spend more in the long term	$1.0	11	11.00
Lost utility revenues	Utility revenues decrease because of energy savings	$−0.8	3	−2.40
Loan interest	Interest paid to banks on construction loans	$0.3	8	2.40
Net job-years: 20-year total				16.40

Note: Table calculations were adapted from a financial scenario based on typical additional green investment and energy use in an office building, as collected for this study; direct and indirect employment multipliers were based on the IMPLAN economic accounts for the United States (Stillwater, Minn.: Minnesota IMPLAN Group, 2006).

1 Table calculations were adapted from a financial scenario based on typical additional green investment and energy use in a green office building, as collected for this study. Direct and indirect employment multipliers are based on the IMPLAN economic accounts for the United States (2006), Minnesota IMPLAN Group, Stillwater, MN. Accessed October, 2007. The job multiplier is the direct and indirect jobs supported by the spending of $1 million for each category of spending in a given sector of the economy. These multipliers are derived from the actual economic data published by the Minnesota IMPLAN Group (http://www.implan.com/).

at a 7% interest rate over a 20-year period to make improvements. Thus, the architect, the engineer, and the building contractor have work immediately, providing a short-term stimulus for the economy. During the next 20 years, as a result of lower energy and water bills, the occupant has more money to spend.[129] The utilities, however, experience a loss or shift of revenues as consumers move some of their spending away from traditional utility services (i.e., consumers purchase less energy).

When this model is run with inputs from a typical green office building, the net increase in jobs is roughly one-third of a full-time job per year. (That is, for each $1 million invested in green building, roughly 20 net job-years are created over 20 years of building life i.e., the equivalent of one job lasting 20 years, when compared with conventional building). A large impact occurs in the first year, due to the initial additional investment in green technologies; smaller but sustained impact in future years is driven by savings in energy costs. Thus, investments in energy efficiency yield increased economic output from increased employment; over 20 years, the value of the employment increase is roughly $1/sf.

While the effects may be small on the level of an individual building, a broad transition to green building can have significant economic impacts. For example, if the United States were to upgrade about one-third of its commercial space to decrease water and energy use by about one-third by 2030, such an effort would create about 100,000 net new long-term, relatively well-paying jobs.

◆　　　◆　　　◆

CONSTRUCTION AND DEMOLITION WASTE

Construction and demolition (C&D) waste from buildings is one of the largest sources of waste generation, and the associated disposal and diversion industries have significant environmental footprints. The EPA estimates that 160 million tons of C&D waste were generated in the United States in 2003, of which only 30%, or 48 million tons, were recycled.[130] In recent years, the C&D waste-recycling industry has grown in many parts of the country, creating new jobs and reducing landfill burden, waste transport costs, and the use of virgin materials in construction.

Green buildings have led in this growing industry. For example, through recycling or reuse, buildings in the study data set divert an average of 79% of C&D waste from disposal. Figure 1.28 shows reported diversion rates in the data set, which range from 50% to almost 100%. As of July 2007, 81% of LEED-NC-certified buildings had achieved a point for recycling at least 50% of their C&D waste, and 59% had achieved a point for recycling more than 75% of their waste.[131] Nationally, it is estimated that 30% of building-related C&D waste is recycled—which means that green buildings divert more than twice as much C&D waste as conventional buildings. Rough estimates based on these figures suggest that if all U.S. construction projects diverted C&D waste at the rate typical of green buildings, an additional 80 million tons of waste would be diverted from landfills eash year.[132] This would be a reduction of over 500 pounds of waste disposal per person per year. (C&D waste recycling may result in additional embodied energy savings if waste transport loads and distances are reduced. As noted earlier, increased use of recycled concrete, steel, and other construction materials generally decreases the energy use embodied in material extraction and production.)

In addition to environmental benefits, C&D waste recycling has significant employment impacts, creating more jobs and associated economic output than disposal does. A number of states have used surveys and economic models to quantify the differences in job creation and economic output from increased recycling. A 2001 report to the California Integrated Waste Management Board (CIWMB) found that for every 1,000 tons of C&D waste, diversion created over 4 job-years, whereas disposal created only 2.5 job-years.[133] Regional analyses, and studies in Arizona, Maine, North Carolina, and Washington comparing diversion to disposal, found between 1 and 10 net additional job-years created per 1,000 tons of waste diverted.[134] Estimates of jobs created directly by diversion only partially capture increases in job creation: industry analyses indicate that roughly two-thirds of the employment impact of recycling comes from the subsequent manufacture and sale of recycled or reused materials.[135]

Even regions with higher baseline rates of diversion offer significant potential for new businesses to process and resell C&D waste. For instance, ReBuilder's Source, a new, South Bronx–based business for salvaging and reselling building materials, is expected to

FIGURE 1.28 Diversion of Construction and Demolition Waste by Buildings in the Data Set

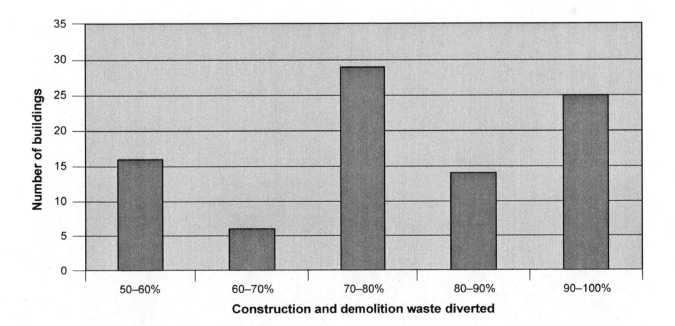

Sally Wilson, *Global Director of Environmental Strategy and Senior Vice President, CB Richard Ellis*

It all starts with corporate social responsibility. When we realized that buildings contribute 40% of the greenhouse-gas (GHG) emissions in the United States, CB Richard Ellis (CBRE) identified GHG emissions as a key component of our social responsibility and decided to take action. CBRE's Environmental Policy, established in May 2007, adopted a goal of carbon neutrality by 2010, and committed CBRE to helping clients engage in more environmentally sustainable practices.

CBRE is the first commercial real estate services company to set a carbon neutrality goal. Carbon neutrality will be achieved through a multifaceted approach in accordance with the guidelines from the World Resources Institute/World Business Council for Sustainable Development Greenhouse Gas Protocol Initiative. This will be used as a baseline for the global reduction of energy usage, the purchase of green energy (where economically possible), and ultimately, the implementation of a robust offsetting strategy that is in alignment with CBRE's global goals and objectives.

CBRE directly manages more than 2 billion square feet of property and corporate facilities globally, so our actions can impact how real estate is built, sourced, occupied, and sold on a large scale. Increasingly, companies that understand that caring for clients and employees is synonymous with caring for the planet often have a competitive advantage.

Particularly in a challenging economic climate, companies need to maintain a balance between fiscal and environmental responsibility. Any sustainable service must include low- or no-cost practices, like implementing recycling programs or helping employees cut energy waste. Clients also need help implementing improvements cost-effectively, whether retrofitting existing space, seeking new space for lease or purchase, or improving existing portfolio operations.

CBRE has embraced a number of existing programs, such as the Environmental Protection Agency's Energy Star program, through which we have assessed energy performance in more than 1,200 buildings, totaling more than 230 million square feet; the Building Owners and Managers Association Energy Efficiency Program, which has trained more than 5,300 attendees on energy-management protocols and techniques; and the U.S. Green Building Councils' Leadership in Energy and Environmental Design Existing Buildings program, in which we are currently evaluating more than 300 buildings for inclusion.[1] CBRE is also embarking on a longitudinal study that will provide valuable metrics—such as annualized client utility savings and long-term value to the asset—and track the success of green efforts over time.

1. CBRE was named EnergyStar Partner of the Year in 2008.

process and sell roughly 400 tons of recycled material per year, while employing the same number of people as a local transfer station that processes 100 times the volume of waste for disposal; the jobs come predominantly from the creation of a value-added product.[136] Many of the jobs associated with C&D diversion and recycling are in low-income areas in need of new sources of employment and economic growth. For all types of recycling, the 2001 CIWMB study estimated a $275 increase in economic output per ton of waste that is recycled instead of being disposed.

Estimating the value of increased employment from C&D waste recycling is difficult, given the limited national data on the industry. Additional research on C&D waste-related industries would help policy makers better understand the employment impacts of recycling in green buildings.

1.10. Property Value Impacts of Building Green

Question: Do green design and construction increase property value?

Evidence: A small number of buildings in the data set gave anecdotal reports of increased property values attributable to greening. A recent analysis showed higher occupancy, rents, and sales prices in green buildings when compared with nongreen buildings. Opinion surveys of architects, owners, engineers, and contractors also generally indicate an expectation that green buildings will enjoy higher property values.

Bottom line: Green buildings appear to have higher property values than conventional buildings.

Many early green buildings were owner-occupied (e.g., schools, government facilities, corporate headquarters), and these owners benefited directly from energy and water savings. However, because these property types are generally not sold, they are not priced on the market, which limits the available information on the value of green buildings in comparison with conventional buildings. A 2005 McGraw-Hill survey of 417 architects, owners, engineers, and contractors found that respondents expected, on average, 7.5% higher property values for green buildings compared with conventional buildings. In 2008, McGraw-Hill's survey found a 10.9% expected increase in green building value, along with an expected 9.9% increase in return on investment (ROI).[137]

Until recently, there was little available data on actual sales prices, rents, or other market indicators of the value of green versus conventional buildings. In a sample of 355 LEED buildings, a 2008 CoStar analysis of office-building sales transactions found that a $24/sf increase in sales price could be attributed to LEED certification.[138] This finding was in line with real estate outlooks published in 2006 and 2007. For instance, Ernst & Young's "Real Estate Market Outlook 2007" noted that "If you aren't at least meeting LEED standards in new construction, there's an increasing risk—one likely to accelerate in the next five years—that your project may falter. Most cutting-edge developments in the years ahead will . . . look to exceed LEED—not just meet it."[139]

In the current real estate downturn, however, the outlook for all types of real estate is in question. As the real estate community becomes more familiar with assessing the value of green buildings, an obvious question has been whether green buildings will be increasingly valued in the down market, or whether owners will shy away from features viewed as unnecessary. Data on annual LEED registrations show continued growth, but a distinct slowing in the rate of growth between the beginning of 2008 and the beginning of 2009 indicates that green building is being affected by larger market trends.

SOURCES OF HIGHER PROPERTY VALUES FOR GREEN BUILDINGS

Higher property values for green buildings come from two principal sources: net operating income; and rent, sales prices, absorption, and occupancy.

Net Operating Income

Net operating income (NOI) is a standard basis for calculating building value. Green buildings may enjoy higher NOI than conventional buildings because of reduced utility costs and lower operations and maintenance costs; further increases in NOI may come about through the potential for lower insurance costs and lower churn costs (the costs of moving within a building), because of greater space flexibility.

To the extent that utility costs are borne by property owners, reduced utility costs directly increase NOI. Calculating a theoretical NPV for green investments involves subtracting the cost of energy-efficiency measures from the present value of future savings over the expected holding period for a property. If lower utility bills are assumed to be permanent, increased NOI can potentially drive substantial increases in building value.

There is some anecdotal evidence to support the view that green buildings are more valuable. One and Two Potomac Yard, LEED Gold offices in Arlington, Virginia, reported a 2% increase in property value on the basis of reduced operating expenses alone.[140] Similarly, on the basis of projected lifetime energy savings, owner Gary Christensen anticipates a $1.5 million increase in the value of the Banner Bank Building, a LEED Platinum building in Boise, Idaho.[141] In describing the energy-savings potential at the National Business Park 318, in Annapolis Junction, Maryland, Peter Garver, of Corporate Offices Property Trust, commented that after energy savings, "other green benefits such as reducing our liability exposure for things like mold and sick-building syndrome are free added benefits."[142]

Reduced insurance premiums are another potential source of increased NOI in green buildings. A 2003 analysis found that 77% of LEED credits are associated with decreased risk of systems malfunctions, which could reduce insurance claims in green buildings. However, insurance companies have been slow to recognize these benefits.[143] In late 2006, Fireman's Fund Insurance Company became the first insurance firm to recognize the lower insurance risks associated with building green by introducing a 5% discount on casualty insurance for LEED-certified buildings. According to the Fireman's, company research revealed that "a green building would be a better building to insure," primarily because of the LEED requirement for building commissioning.[144] The leading causes of insurance claims are electrical problems and water damage; commissioning acts as an extra quality-control check on building systems. Fireman's research found that green building systems should be expected to malfunction less frequently, generating fewer insurance claims. As of December 2007, about ten green commercial buildings had enrolled in Fireman's discount program. Fireman's appears to be the only insurance company currently offering such a discount.

In 2008, Fireman's introduced a more popular green upgrade program, under which building owners pay a small premium to ensure that rebuilding and renovation will be LEED-certified in the case of any damage, including total loss. Similarly, California-based New Resource Bank offers lower lending rates for new green construction, providing an additional reduction in costs to owners.[145] Because of higher NOI, green building owners are considered less likely to default. And, should default occur, the higher value of a green building lowers the bank's risk of loss.[146]

Rent, Sales Prices, Absorption, and Occupancy

Reports from recent green buildings indicate a general trend toward higher rents and sales prices. A 1999 survey of 1,800 office workers by the Building Owners and Managers Association International and the Urban Land Institute indicated that IEQ is one of the most important components of workplace satisfaction—a preference that may underlie the higher value of green buildings. Survey respondents assigned the highest importance to features related to tenant comfort, including air temperature (95%) and IEQ (94%). Office temperature and the ability to control it were the only features that were rated "most important" and that were categorized among the characteristics with which occupants were least satisfied in typical office buildings.

Recent reports of the impact of green buildings on rents or sales prices include the following:[147]

- The Solaire, a LEED Gold residential building in Battery Park City, Manhattan, experienced a 10% to 15% sales premium per unit when compared with similar nongreen buildings in the area.[148]
- Two green buildings developed by the Liberty Property Trust in Philadelphia— the Comcast building and One Crescent Drive, in the Philadelphia Navy Yard— have sustained rents that are 25% to 50% higher than market rates. According to Liberty's senior vice president, John Gattuso, One Crescent Drive, a LEED Platinum building, was valued at almost 90% more per square foot than any

other building in Philadelphia. Gattuso attributes this difference in large part to the building's green design and certification.[149]

- The Louisa, a green multifamily building in Portland's Brewery Blocks redevelopment, was nearly 70% leased within a few months of opening—exceeding the speed of absorption for comparable space by four months. The property manager notes that the building has maintained very high occupancy rates and low turnover rates, and that the perception of a "healthy living environment" seems to be the most significant attraction for tenants—even greater than the perceived value of reduced utility costs.[150]

Given the higher values associated with green properties, new firms have recently emerged that are dedicated to retrofitting and repositioning existing conventional buildings as green buildings. Several real estate funds, including Thomas Properties, and Canyon-Johnson Urban Communities Fund, are focused on greening multiunit residential properties in urban areas.[151]

INDUSTRY SURVEYS AND ANALYSIS

Many developers and real estate appraisers are closely watching the market—unsure, as yet, whether green buildings will, in the long run, command higher property values. How will green buildings be perceived in a slower real estate and construction market? Will green features be seen as an unnecessary frivolity, or will there be a migration to green quality?

Recent analysis by CoStar, which maintains a database with information on 40 billion square feet of building space, suggests that green buildings have positive impacts on rents, occupancy, and sales prices.[152] Moreover, these impacts appear to be larger than would be expected from energy savings alone—and larger than anticipated by industry opinion surveys.[153] In 2006, CoStar began adding LEED certification information to its database. Energy Star ratings, used to assess energy performance in existing buildings, were added to the database in 2007, along with a feature that allows users to search for buildings by LEED or Energy Star rating. On CoStar's online database, LEED and Energy Star buildings now automatically appear toward the top of search results.

A 2008 CoStar analysis compared the value of LEED and Energy Star buildings to nongreen buildings that were matched on the basis of size, age, class, and submarket. The analysis pool included 973 Energy Star and 355 LEED buildings.[154] Table 1.6 summarizes the results of the comparison during the first quarter of 2008.

Nationally, LEED and Energy Star offices showed significantly higher occupancy rates, rents, and sales price when compared with similar nongreen offices. The CoStar study has several limitations, however: first, because of the small sample size, CoStar could not fully compare LEED buildings with nongreen buildings in the same submarkets;

TABLE 1.6 PROPERTY CHARACTERISTICS OF LEED-CERTIFIED, ENERGY STAR–CERTIFIED, AND NONCERTIFIED OFFICES,
FIRST QUARTER OF 2008

	LEED-certified offices	Non-LEED offices	Difference	Percent difference
Occupancy rate	92%	88%	4%	5%
Rent (/sf)	$42	$31	$11	35%
Property value (/sf)	$438	$267	$171	64%
	Energy Star offices	Non-Energy Star offices	Difference	Percent difference
Occupancy rate	92%	88%	4%	5%
Rent (/sf)	$31	$28	$3	11%
Sales price (/sf)	$288	$227	$61	27%

Source: J. Spivey, "Commercial Real Estate and the Environment," CoStar, 2008 (www.costar.com/news/Article.aspx?id=D968F1E0DCF73712B03A099E0E99C679).

second, a limited number of sales transactions were available for comparison.[155] Real estate appraisers may also question the large-scale statistical approach because it is less detailed and focused than the normal valuation process for commercial properties. Despite these limitations, the CoStar analysis is noteworthy as the first published comparison macro market data for sets of green and nongreen buildings across the country.

Along with the peer comparisons, CoStar used a statistical model to remove the effects of age, location, size, and other factors on the variations in sales price. In this analysis, CoStar estimated that LEED certification had increased sales prices by $24/sf, which is less than the $171/sf increase shown in table 1.6. The $24/sf increase in sales price is roughly 2.5 times the value of energy savings alone. Unlike LEED, the Energy Star standards require only energy efficiency in buildings; the substantially higher disparity in sales prices for LEED versus Energy Star buildings ($171/sf versus $61/sf, as shown in table 1.6) indicates that consumers place substantial value on the many benefits of LEED that go beyond energy efficiency.

Differences in occupancy rates, rents, and sales prices between Energy Star and non-Energy Star buildings varied over the study period and between different locations; generally, however, the relative value of LEED and Energy Star buildings versus nongreen buildings increased between 2006 and 2008. A rise in the relative value of green buildings is consistent with a trend of accelerating interest in, and awareness of, green design.[156]

A 2008 McGraw-Hill survey of architects, owners, engineers, and contractors found that a large majority (77%) anticipated increased sales growth from green properties. On average, respondents expected to see a 10.9% increase in building value, a 9.9% increase in ROI, a 6.4% increase in occupancy, and a 6.1% increase in rent—all representing significant increases in perceived value in comparison with the 2005 version of the same survey.[157] However, those surveyed tended to be more familiar with, and active in, green building than typical members of their industries, suggesting caution in applying these findings generally.

Similarly, in homebuyer surveys, a majority of respondents say that they are willing to pay more for green homes. In a 2005 survey by Chistopherson Homes, 50% of respondents reported that they would be willing to pay an additional $100/month in mortgage payments for a green home—echoing the results of similar surveys in California, Denver, Salt Lake City, and San Diego.[158] In the spring of 2007, the National Association of Home Builders released an assessment of the demand for green housing in which buyers indicated that they would pay a 5.8% premium for a green home.[159] In a consumer survey in the Northwest, 78% of respondents said that they would give preference to a home on the basis of energy efficiency, and a majority indicated that they would be willing to pay a premium for energy-saving systems and design.[160] It should be noted that preferences expressed in surveys do not necessarily translate into actual consumer spending. These survey results should therefore be taken as evidence of a general preference for green homes, rather than as an indication of the premium buyers would actually pay for green homes.

The surveys just described were conducted before the current real estate downturn; the impact of the current market on green building remains uncertain. Some recent surveys suggest that green buildings may be faring better than conventional buildings in the down market. According to McGraw-Hill's *Green Outlook 2009*, 40% of real estate professionals believe that green homes are easier to market than conventional homes in the down economy, while 29% believe that they are more difficult to market.[161] Among consumers, 70% report being more inclined to buy a green home in the down economy—perhaps because of an expectation that green homes will have lower operating costs. (For a discussion of green development in the market downturn, see "Perspective: Investing in Brownfields.")

In the case of commercial buildings, however, surveys show that many respondents believe that green buildings will be hit harder by the downturn than conventional buildings. Of the building professionals surveyed for McGraw-Hill's *Green Outlook 2009*, 40% believe that green buildings will be more affected by the current market than conventional buildings, and 25% believe that conventional buildings will be more affected. Among USGBC members, however, opinions are reversed, with 32% believing green buildings will be less affected and roughly 20% believing that they will be more affected.[162] Time will tell whether the greater confidence of USGBC members reflects greater experience with green projects in the marketplace, or an aspiration that the larger market will increasingly value green buildings as USGBC members do.

The net impact of green buildings on property value includes the additional investment

An Interview with Tom Darden, Chief Executive Officer, Cherokee Investment Partners

In 1984, Tom Darden acquired a brick business.[1] This allowed him to be an early innovator of environmentally friendly practices, such as the use of carbon-neutral biomass to power the brick plants, and the development of methods to clean up petroleum-contaminated soil. Building on this experience, Tom Darden and his business partner, John Mazzarino, founded a new company to focus exclusively on environmentally impaired real estate; in 1996, Tom and John started their first investment fund to buy brownfields (contaminated sites), clean them up, and sell them for redevelopment.

Today, Cherokee Investment Partners, the company that Darden and Mazzarino founded, is investing in its fourth fund, valued at $1.2 billion, and has grown into the leading private equity firm investing in the redevelopment of infill properties. In 2006, Cherokee created and adopted a set of sustainability criteria to help guide its investment decisions. Tom Darden, Cherokee's chief executive officer, shared his perspective on the financial value of this green focus.

Cherokee's business model has been driven, in large part, by a philosophy that financial, environmental, and social performance are inextricably linked. We didn't have a sustainability mission in mind when we started, although I believe we were practicing sustainability as we worked to bring capital to clean up and redevelop brownfield sites.

We built the National Mainstream GreenHome in Raleigh, North Carolina, to examine various green building methods and materials and to share what we are learning with others. The home combines materials that protect human health and the environment with mainstream aesthetics and comfort. It was the first LEED for Homes Platinum project in the Southeast. The GreenHome is expected to achieve water savings of 50% to 60% compared to a conventional home. With a ground source heat pump, a tight building envelope, solar photovoltaic and thermal systems, and North Carolina–based green energy credits, we hope to nearly eliminate the home's energy and CO_2 footprint.

In 2007, Cherokee invested in a green headquarters building in Raleigh. As a LEED for Commercial Interiors Platinum project, our office is one of only a few historic restoration projects in the world to achieve this designation. Increasingly, we believe that employees want an office that is healthy and respectful of the environment. This affects us in a very direct sense: we are able to attract talented people who want to be in this kind of environment. The money spent on a green building could be offset by one good hire. That said,

we didn't spend much more on green features in our office. In fact, per-square-foot costs for the upfit fell within the normal range for comparable upfits in the market. Cherokee also takes part in the LEED for Neighborhood Development pilot program. One Cherokee project in the pilot program is Faubourg Boisbriand. Located in the Montreal metropolitan region, the former General Motors factory is being redeveloped as a mixed-use community with a variety of housing types, shops, a community civic center, and a possible rail stop. The project also has community parks, pedestrian walkways, and jogging and bike paths. Cherokee's development partners are planting 2,400 trees and reusing 140,000 tons of concrete and 2.5 million square feet of asphalt. As a result of the redevelopment, local tax revenue will also increase. The community recently earned Gold certification in the LEED-ND program, one of only a handful of pilot projects across North America to achieve this designation.

Are there additional costs associated with this type of green community-scale project?

My initial answer would be no—there could well be savings if you can reduce infrastructure costs and narrow streets, but our ability to control these aspects of development can be limited by planning codes and ordinances, site context, and the market. Generally, green community design is cost neutral and/or revenue positive where there is a market that is willing and able to accept greater density.

We have found that there are limits to the application of sustainability principles. When costs for green strategies exceed an acceptable level, there is a natural push-back from developers, from fund managers like us, and from our investors. As costs decrease over time, the ability to invest in and build high-performance projects at a scale that can make a real difference will come into focus.

Traditionally, Cherokee would invest in land and in the environmental remediation of that land. Once the environmental work was finished, we would exit the deal by selling it to a vertical developer. Increasingly, we have been motivated to affect the form and type of vertical development after remediation. This helps us capture economic value in later stages of the development process and, I believe, generate increased value for our investors.

(continued on page 80)

1. Greg Kats serves on the sustainability advisory board for Cherokee.

Is there a risk in not building green in the current market?

From my perspective, the risk is to miss a growing market by not investing in green development. Of course, all portfolio managers also need to assess whether they have assets that will be at risk in a carbon-constrained world. In addition, approximately 190 state and local governments now require new buildings to achieve some level of green certification. In other words, the green building barrier to entry has already been set in certain markets.

What is the future of green development, given the state of the economy?

Naturally, in a severely constrained capital market, we are focused on achieving strong economic returns for our investors—and we aim to do that, in part, by advancing environmentally responsible development.

Despite today's sober economic news, the world has crossed a threshold such that in the years to come, businesses that conduct themselves in a sustainable and responsible manner will be positioned to succeed. It is clear that we are going to have to work together to make it happen on a scale that will make a measurable difference.

(the green premium), operational savings (higher NOI), and higher sales prices. Figure 1.29 illustrates this impact: the cost premiums and energy savings were drawn from office buildings in the data set; the calculations assume a sale in year 5 that achieves the $24/sf increase in sales price indicated for LEED office buildings in the CoStar study. In this example, the financial impacts of greening are as follows:

- Green premium: average $3.50/sf
- Operational energy savings: average annual savings of $0.57/sf in year 1, with 5% escalation in annual energy savings in subsequent years. (This is consistent with a 34% reduction in average commercial building energy expenditures in CBECS, adjusted to 2008; see section 1.3, "Energy-Use Reductions.")
- In a sales transaction that occurs between year 5 and year 6, a sales price that is $24/sf higher than the price of a conventional building.

RISK

Green buildings reduce the risk of future energy price increases, and can reduce operations and maintenance costs and potential liability from health problems caused or exacerbated by poor IEQ. With CO_2 regulations expected under the Obama administration, new green construction and the greening of existing portfolios may reduce carbon risk for building owners. Many major corporations are publicly reporting CO_2 emissions, even in advance of regulation. In February 2008, over 50 institutional investors from the United States and Europe pledged to invest $10 billion over the next two years in technologies to reduce CO_2 emissions, and to reduce energy use in their real estate portfolios by 20% over the next three years.[163] Similarly, the Carbon Disclosure Project (CDP) is a voluntary international effort to promote reductions in CO_2 and to collect information on the carbon emissions and climate-change mitigation strategies of leading corporations. As of 2008, the CDP had 385 signatory investors managing over $57 trillion in assets.[164]

But for many in the construction and real estate industries, it is the value of keeping up—that is, reducing the risk of obsolescence in nongreen buildings—that is perceived as perhaps the largest risk-reduction benefit of green building. According to Marty Dettling, vice president of the Albanese Organization (which developed The Solaire, a LEED Gold multifamily project in New York City):

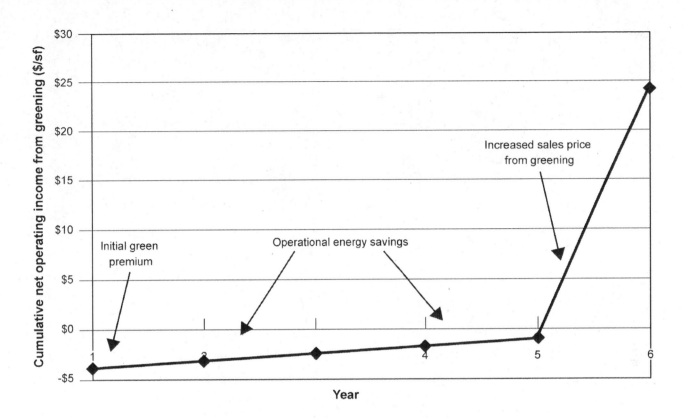

FIGURE 1.29 Impact on Property Value of Energy Efficiency in Green Buildings: Illustrative Graph

Albanese Organization . . . realizes that the current consensus in the industry is that Class A building construction requires that the building also include sustainability measures in its design. Those buildings that do not include such measures will find that in a short time that they do not contain what the market demands for a healthier living, working and more energy-efficient environment.[165]

Echoing this sentiment, Ernst & Young's "Real Estate Market Outlook 2007" noted that "green building—once dismissed by major developers as 'too expensive'—becomes almost a necessity as tenants, lenders, residents, and even investors push for sustainability. . . . Look for green principles to become synonymous in the real estate industry with solid, cost-efficient operating principles."[166]

In 2007, Vance Voss, of Principal Real Estate Investors, commented on the risk of obsolescence in nongreen buildings:

In the $560 million of green projects we have been involved in, we definitely think there is a return on the incremental cost for doing those green deals. We believe it will be an increasing benefit going forward that green projects will be open to a broader spectrum of tenant demand and reduce the potential obsolescence risk of buildings over the long term.[167]

The rapid growth of green buildings from 2000 to 2007 supports the notion that green practices and certification have the potential to become the construction industry standard. In recent years, at least 27 states and 103 cities, towns, and counties have passed ini-

Note: Data through year 5 are based on average reported green premium and energy savings from the buildings in the data set. In year 5, it is assumed that the asset will be sold, realizing an increased value consistent with CoStar findings. See J. Spivey, "Commercial Real Estate and the Environment," CoStar, 2008 (www.costar.com/news/Article .aspx?id=D968F1E0DCF73712B03A 099E0E99C679).

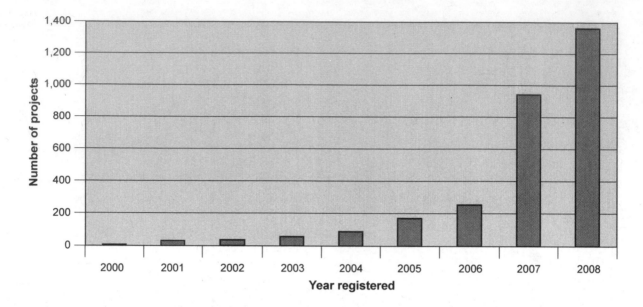

FIGURE 1.30 Growth of
LEED-Registered Projects
(New Construction and
Core and Shell)

tiatives requiring or encouraging new construction to be green.[168] According to McGraw-Hill Construction, the percentage of surveyed corporations reporting a "significant share" of green building more than tripled between 2003 and 2008, to 37%.[169]

The square footage of LEED-registered building space has grown steadily since 2000, with 1.3 billion square feet of space registered for LEED-NC and LEED Core and Shell in 2008 alone. The rate of growth has fluctuated over time, with a compound annual growth rate of 69% between 2002 and 2008 (see figure 1.30).[170] Annual growth in LEED registrations depends not only on the rate of adoption of green building, but also on the larger construction market, and on changes in the standard itself. Scheduled changes to the standard can spur a rush of registrations before new rules come into effect. The introduction of LEED for Neighborhood Development (LEED-ND), in 2007, allowed projects of a larger scale to participate, significantly expanding the total number of registered projects beyond the numbers in figure 1.30.

However, beginning in the second half of 2008 and continuing into 2009, LEED registrations experienced a distinct slowdown: USGBC records show a 52% increase in LEED square footage registered between the first and second quarter of 2008, declining steadily to a 48% *decrease* in new registered square footage between the last quarter of 2008 and the first two months of 2009. Without data on total U.S. construction starts during the same period, we cannot say whether the percentage of U.S. construction that is green is changing, but it is clear that green building is being affected by the larger economic downturn. Whether green design and efficiency become the norm largely depends on the mainstream market understandng of the cost effectiveness of greening.

Building green appears to create higher property values than conventional, inefficient design and construction, through higher NOI, rents, sales price, and occupancy rates, and through reduced risk. The magnitude of these impacts is difficult to estimate, however, and remains an open question in the current market downturn. Ultimately, the perception

Shyam Kannan, Robert Charles Lesser & Company

In the spring of 2007, Robert Charles Lesser & Company (RCLCO) conducted a national survey of homeowners to gain an understanding of their attitudes toward green residential products. The survey was designed not only to assess demand for green homes, but also to determine what characteristics were associated with green homes, and to gauge the importance of green attributes in comparison with other factors influencing home choice. Almost 10,000 respondents received the surveys, generating over 1,000 complete responses.

Through a survey strategy that asked respondents to make a series of sequential trade-offs between 42 factors (from type of residential product to school quality, lawn size, and ceiling height), RCLCO sought to identify the most important factor in respondents' next home purchase. Thirty-six percent of respondents ranked a "green" factor—the environment, energy savings, or health benefits—as their top priority in home choice, outweighing factors such as location and school quality.[1] RCLCO believes that the larger group of buyers who find these motivating factors important, but not all-important, represent the "convincible" market—those who, through additional education or aggressive marketing, can be persuaded to buy a green home. For these buyers, green may be only a tie-breaking decision among otherwise equal housing options.

More detailed survey results suggest that there may be a gap between the actual benefits of green homes and consumer knowledge of these benefits. For instance, while only 6% of homebuyers consider "the environment" the top priority in home choice, almost 75% of homebuyers do not believe that their home has an adverse environmental impact at all.

Demonstrating the cost-effectiveness of green buildings will be influential for this market segment. As can be seen in figure 1, assuming the investments would pay them back over time, a large percentage of respondents reported that they were willing to pay more for a home if it will provide energy savings or health benefits, or is good for the environment.

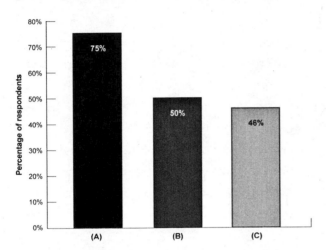

FIGURE 1 Assuming That Their Investment Pays Them Back Over Time, Percentage of Respondents Indicating That They Would Be Willing to Spend Additional Money If Their Home (a) Saves Energy, (b) Provides Health Benefits, (c) Is Good For the Environment

---

1. RCLCO, "Green Residential Development." Among the characteristics that respondents had to rank were type of residential product and neighborhood, excellent school systems, large lawns, luxury finishes, and high ceilings.

---

of the cost-effectiveness of green building—including first costs, energy savings, health impacts, and other factors—will determine what the market is willing to pay for green buildings. Greater availability of data on the impacts of green building will allow real estate professionals, homebuyers, and renters to more accurately gauge the desirability—and value—of green buildings, and would influence the level of investment in greening.

1.11. Net Financial Impacts of Green Buildings for Owners and Occupants

In this section, we use simple payback and NPV analysis to calculate net financial impacts for each building in the data set. We compare the full range of costs and benefits for a typical green office and a typical green school, both based on the median-case benefits estimated in previous sections of this book, including public and private benefits. Comparing costs and benefits involves choosing the range of benefits to consider, selecting a period over which to compare benefits, and making assumptions about the future escalation of costs and benefits.

This section focuses primarily on costs and benefits to building owners and occupants. Policy makers and community advocates interested in evaluating the public benefits of green building can perform similar analyses.

MEDIAN CASE: GREEN SCHOOLS AND OFFICES

Figure 1.31 summarizes the costs and benefits for green schools and green offices using (1) the median reductions in energy and water use reported in the data set and (2) baseline expenditures for typical schools and offices. Present-value benefits are calculated over 20 years, with a 7% discount rate and a 5% annual increase in energy and water costs. The figure includes only the most easy-to-quantify benefits, and those for which data were supplied for a majority of the buildings in the data set: the green premium, energy savings, and water savings. When only energy and water benefits are included, the NPV (total benefits minus the green premium) is roughly $5/sf for a typical green school and $7/sf for a typical green office. This figure excludes benefits that relate to health, productivity, and property value; it also excludes societal benefits such as job creation and emissions reductions.

Figure 1.32 shows illustrative productivity benefits and public benefits, including job creation and emissions reductions. Adding productivity and public benefits to the picture increases the total public and private NPV to roughly $21/sf for green schools and $24/sf for green offices. In the case of green schools, the productivity impacts would ultimately take the form of improved student performance and long-term earnings.[171] The value of CO_2 emissions is based on an assumed price of $15 to $20 a ton, increasing at 2% annual inflation over 20 years (see "Valuing CO_2 Reductions," in section 1.3).

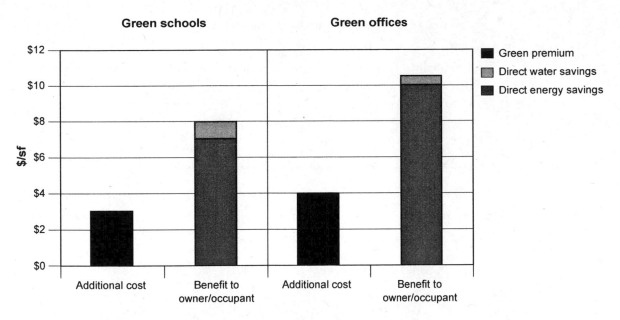

FIGURE 1.31 Present Value of 20 Years of Costs and Benefits of Energy and Water Savings for Green Schools and Green Offices in the Data Set

Note: Additional benefits not estimated include health, productivity, student performance, impacts on property values, indirect impacts on water systems, brand benefits, savings on operations and maintenance, and embodied energy savings.

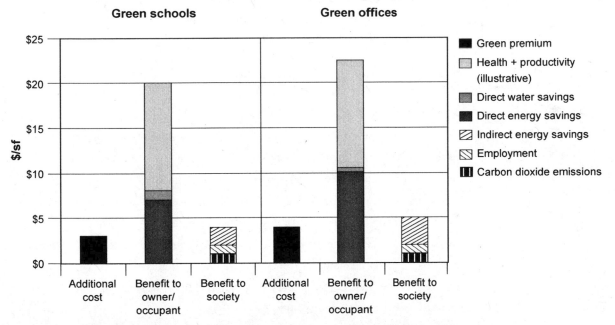

FIGURE 1.32 Illustrative Present Value of 20 Years of Private and Public Benefits for Green Schools and Green Offices in the Data Set

Notes: (1) Additional benefits not estimated include impacts on property values, indirect impacts on water systems, brand benefits, savings on operations and maintenance, and embodied energy savings. (2) The graph is based on findings for the study set and on the literature. There is significantly greater uncertainty, and less consensus, surrounding methodologies for estimating health and societal benefits.

Both graphs exclude a range of significant but difficult-to-quantify benefits, including impacts on property values, student performance, brand benefits, and embodied energy; indirect impacts on water systems; reduced storm-water flow; savings on operations and maintenance; and mitigation of climate change (apart from what is reflected in the CO_2 price of $15 to $20 a ton).

COST-EFFECTIVENESS OF GREEN BUILDINGS

To analyze the cost-effectiveness of the green buildings in the data set, we performed simple payback and NPV calculations for individual buildings. Each method of measuring cost-effectiveness was performed for three scenarios:

- Scenario A includes only up-front cost premiums and annual energy and water savings.
- Scenario B includes cost premiums, energy and water savings, and an illustrative estimate of health benefits, based on potential reductions in respiratory infections and allergies.
- Scenario C includes cost, energy and water savings, and an illustrative estimate of productivity benefits based on significant reductions in sick-building syndrome and in the associated loss of productivity and work time. (For a discussion of health and productivity benefits estimates, see "Estimating Health and Productivity Benefits," in section 1.7).

Simple Payback

Simple payback is calculated by dividing the initial cost (the green premium) by the expected annual benefits, without applying a discount rate to account for the time-value of money. The result is the time it takes to pay back an initial investment in green building, after which all future savings accrue as financial gains for owners and/or occupants. Thus, for green buildings that report a 0% cost premium, simple payback is immediate, and all savings are net financial benefits. For this section of the report, we calculated simple paybacks of the green premium for scenarios A, B, and C. Increased property value, in the form of increased rents or sales price, was excluded from these calculations and could shorten payback periods significantly.

For the 103 buildings in the data set that reported green premiums and energy savings in dollars per square foot, we calculated simple paybacks from energy savings for the reported green premiums. For buildings that reported reductions in energy and water use only in the form of percentages, we based dollars-per-square-foot savings estimates on typical baseline expenditures by building type. Median simple paybacks in the data set

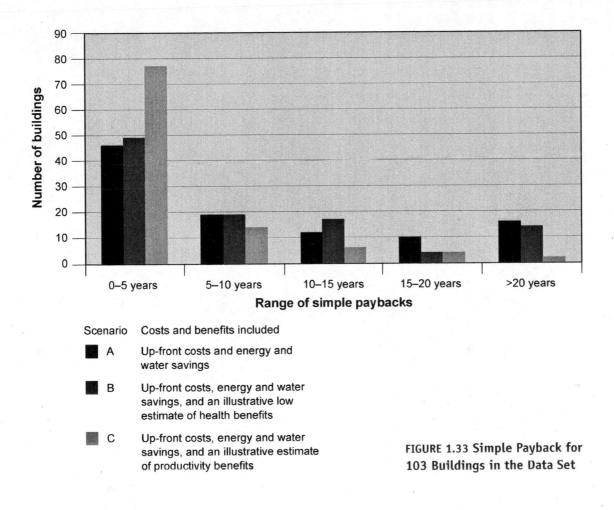

Scenario — Costs and benefits included

■ A — Up-front costs and energy and water savings

■ B — Up-front costs, energy and water savings, and an illustrative low estimate of health benefits

■ C — Up-front costs, energy and water savings, and an illustrative estimate of productivity benefits

FIGURE 1.33 Simple Payback for 103 Buildings in the Data Set

were 6.2 years, 5.6 years, and 2.4 years for scenarios A, B, and C, respectively. Figure 1.33 shows the distribution of simple paybacks for each scenario.

As can be seen in figure 1.34, the large majority (about 80%) of buildings in the data set show positive NPVs from energy and water savings alone. Median NPVs are roughly $5/sf, $6/sf, and $17/sf for scenarios A, B, and C, respectively. Notably, when estimated productivity benefits are included (Scenario C), a majority (57%) of buildings show NPVs greater than $15/sf—suggesting that significant health and productivity benefits, where achievable, could, in effect, subsidize investments in more expensive green building features that have longer paybacks.

Overall, our findings suggest that typical green buildings are cost-effective, based on reductions in energy and water use alone; including a fuller range of benefits makes the financial case for green buildings even stronger. Individual green buildings experience a wide range of cost-effectiveness, depending on the particular building, the level of energy and water savings, and the particular benefits included in the analysis. Green buildings in the data set range from projects with long payback periods (more than 20 years), to proj-

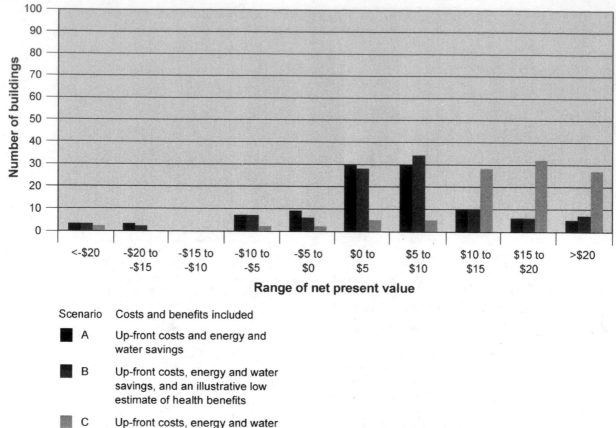

Scenario | Costs and benefits included

■ A   Up-front costs and energy and water savings

■ B   Up-front costs, energy and water savings, and an illustrative low estimate of health benefits

■ C   Up-front costs, energy and water savings, and an illustrative estimate of productivity benefits

FIGURE 1.34 Net Present Value of 20 Years of Costs and Benefits for Buildings in the Data Set

ects with immediate paybacks and NPVs in excess of $20/sf. In the conservative scenario (including only energy and water savings), half the buildings in the data set show simple paybacks of six years or less, and NPVs greater than $5/sf over 20 years. In the fuller benefits scenario, including health and productivity benefits (equivalent to less than 0.5% improvement in productivity), 75% of buildings in the data set show simple paybacks of five years or less, and 84% show NPVs greater than $10/sf over 20 years. All three scenarios exclude substantial and real additional benefits, and therefore tend to underestimate the financial benefits of greening.

PART II

COSTS AND BENEFITS OF GREEN COMMUNITY DESIGN

The green strategies and impacts discussed so far in this book
relate to the design and construction of individual buildings.
While about 45% of the energy used in the United States is
consumed in buildings, roughly 28% is consumed by
transportation—primarily by the task of moving people between
buildings.[1] Green community design—that is, compact,
walkable, appropriately situated development that incorporates
green buildings—can provide energy savings and health benefits
far beyond what green buildings alone can offer. Indeed,
achieving deep reductions in CO_2 emissions, as envisioned by
the Obama administration, is probably not possible without
widespread implementation of both green building and green
community design strategies. The health effects of obesity, for
example, are estimated to cost the United States over $100 bil-
lion annually in medical costs and lost work, and obesity is
strongly associated with declining physical activity in car-dominated, conventionally sprawling communities.

The intent of part 2 is to help build a framework for
comparing the costs and benefits of green and conventionally
designed developments for all involved groups: developers,
owners, residents, and the public. Knowledge of many
disciplines (transportation, health, hydrology, ecology, traffic,
sociology, demographics, public infrastructure planning, and

real estate market analysis) is needed to fully understand the costs and benefits of community design. What we provide here are preliminary estimates of the magnitude of the costs and benefits, on both a dollars-per-household and dollars-per-square-foot basis, by highlighting and synthesizing recent research and case studies. Additional research is necessary to better understand the costs and benefits of green development and community design as alternatives to conventional development.

Before World War II, most communities were composed of compact neighborhoods in which many residents could satisfy most of their needs, which commonly included being able to reach school, work, and shopping on foot or by means of public transportation.[2] Over the past 60 years, conventional development has become dominated by cars—making walking, biking, and the use of public transport increasingly difficult (and, in the case of walking and biking, often unsafe). Zoning rules have mandated the segregation of homes from schools, offices, and commercial districts, all too often preventing people from using any form of transportation other than cars to meet their daily needs. Street networks are characterized by wide arterial roads that link disconnected pods of residential development with shopping centers, schools, and offices—all of which are set back from the main road and surrounded by parking lots. For the last two generations, most new developments have consisted of communities where driving is generally the only practical way to get to work or school, shop for basic necessities, see friends, or participate in recreational or civic activities.

This type of low-density suburban development, sometimes referred to as sprawl, typically occurs on greenfield sites, and has, until recently, been assumed to be both cheaper and more profitable for developers than mixed-use, urban-infill, or walkable projects. It has also often been touted as the preferred housing type for homebuyers. From the 1960s through the 1980s, economic decline in the center of many U.S. cities encouraged—

and was exacerbated by—a rapid migration of middle-class families out of cities and into the sprawling new suburbs. Massive and sustained public investment in new roads, and in water, sewer, and utility infrastructure, effectively subsidized and fostered this type of development. At the same time, local infrastructure, services, and public transportation were chronically underfunded in cities.

The cost of sprawl is well documented. In 2001, for example, the Texas Transportation Institute estimated that traffic congestion, which is exacerbated by conventional, sprawling development, cost the nation $69.5 billion in lost productivity and wasted fuel.[3] People living in denser, more walkable neighborhoods, with a mix of uses, generally get more exercise and have a lower incidence of obesity, leading to reduced costs for the health problems associated with physical inactivity.[4] Potential savings for U.S. local governments, in terms of reduced costs for roads and water and sewer infrastructure under compact development, have been estimated at 11% of infrastructure spending, or $126 billion over 25 years.[5] Widespread adoption of smart growth policies nationwide could save an estimated 1.5 million acres of agricultural land and 1.5 million acres of ecologically sensitive land by 2025.[6]

Increasingly, community groups oppose new conventional development because of the resulting increase in traffic congestion, the strain on municipal infrastructure and budgets, and concerns about health issues and loss of sense of community. Green community design has emerged as a positive alternative to simply opposing development: green communities can be designed and built to provide a range of environmental, social, and economic benefits.

While green developments may incur additional first costs to developers or increase design time and complexity, both existing research and analyses undertaken for this book indicate that the benefits to owners and the public, and the higher revenues for developers, typically outweigh the potential additional first costs. Moreover, green neighborhood development can sometimes yield significant first-cost savings: for example, analysis of ten projects encompassing over 1,500 homes demonstrates that ecological conservation development can reduce initial infrastructure costs by an average of 25%, or $12,000 per home (see section 2.8, "Cost Savings in Ecologically Designed Conservation Developments").

The benefits of green community design commonly include reductions in vehicle use and associated reductions in emissions and gas costs; improved health from increased physical activity; lower costs for road, water, utility, and storm-water infrastructure; increased safety from narrower streets; higher long-term property values; and improved community cohesion and increased civic participation. Preliminary estimates based on research in Atlanta indicate that green community design can reduce transportation costs by $800 per household per year (or up to $4,600 per year, if car ownership decreases). Because of increased physical activity, health costs are estimated to be reduced by $300 per household per year. The present value of 20 years of these health and transportation benefits is on the order of $25,000 per household, or about $10 per residential square foot.[7]

2.1. What Is a Green Community?

Three dimensions determine the greenness of neighborhood and community-scale development projects: location, neighborhood design, and green building. Table 2.1 presents a simplified matrix of the options, ranging from the conventional (inefficient buildings in sprawling communities) to the most green: efficient, healthy buildings in walkable, compact, mixed-use neighborhoods with high-performance green infrastructure. Table 2.2 compares green community design with the conventional sprawl that has dominated community development in the United States for more than a half-century.[8] In contrast to conventional, car-dominated development, green communities do not require residents to drive to reach most or all destinations, but instead provide mobility choice, and are designed to allow people to walk to schools, jobs, stores, restaurants, open space, trails, or other destinations. Green communities typically include a mix of uses, including residential and commercial space, public parks, and open space; they are also characterized by reduced storm-water runoff and green, high-performance infrastructure and buildings.

The most sustainable community design choices include all three dimensions of green community design. Figure 2.1 provides a graphic representation of the relative costs of green community development and conventional sprawl for the owner or resident, the developer, and the public.

TABLE 2.1 DEVELOPMENT OPTIONS MATRIX

	Conventional building	Green building
Conventional neighborhood design	Most recent U.S. buildings and development	Most existing green buildings
Green location and community design	Most existing traditional neighborhood developments, conservation developments, and walkable urban centers	Developments that incorporate both green building and green community design (e.g., LEED for Neighborhood Development projects)

TABLE 2.2 GREEN COMMUNITY DESIGN VERSUS CONVENTIONAL SPRAWL

	Location	Neighborhood design	Building design
Conventional sprawl	Greenfield Car-dominated	Car-dominated Low-density Segregated uses Pedestrian-unfriendly High paving and grading costs Increased storm-water runoff and pollution	Energy- and water-inefficient Poor indoor environmental quality Materials with high embodied energy and environmental impact
Green community	More mobility choices, less car dependence Walkable proximity to destinations (e.g., jobs, stores, schools, parks) Commonly in a brownfield, in an urban infill location, or adjacent to an urban area Public-transit-oriented	More mobility choices, less car dependence Compact, dense, and walkable Greater land use mix and connectivity Defined streets On-site storm-water infiltration Habitat preservation and restoration Public open space	Efficient use of water and energy Healthy for occupants Sustainable materials
Benefits of green communities when compared with conventional sprawl	Reduced congestion Reduced driving Increased open space Reduced development pressure on farms and ecosystems	More opportunities for walking and other physical activity Greater pedestrian safety Reduced storm-water runoff and pollution Reduced public infrastructure costs Improved community life Increased long-term property value More open space	Reduced energy and water use Improved occupant health Higher property value Increased employment Reduced emissions

As is the case with individual buildings, *green* refers to a number of distinct goals. Green developments typically have several of the following goals:

- Increase density, walkability, land use mix, access to public transport, and pedestrian-friendliness
- Reduce required driving and related costs
- Reduce storm-water runoff and pollution
- Preserve and restore open space and habitat
- Increase the energy-, water-, and materials-efficiency of buildings and infrastructure
- Support more diverse, economically sustainable communities (see box 2.1).

Many existing communities, including most built before WWII, that may not self-identify as green have attributes that are green in comparison to the car-oriented sprawl that has dominated new U.S. construction for more than a half-century.

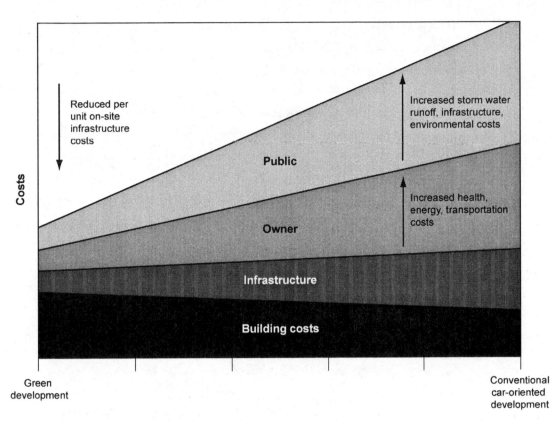

Reduced per
unit on-site
infrastructure
costs

Increased storm water
runoff, infrastructure,
environmental costs

Public

Increased health,
energy, transportation
costs

Owner

Infrastructure

Building costs

Costs

Green
development

Conventional
car-oriented
development

FIGURE 2.1 Illustrative Costs of Green Community Design versus Conventional Development

BOX 2.1 TYPES OF GREEN COMMUNITIES

Ecological conservation development is an approach to residential design that enhances views and access to open space by clustering homes on smaller lots and conserving a large protion of natural space.[i]

Mixed-use, high-density development refers to projects that combine residential units, multistory offices, and retail or other uses; such projects often include multiple buildings in an infill setting.

Smart growth is a broad term that encompasses a range of development types that accomplish the following environmental objectives: reduction in water and energy use; increased walkability and use of public transport; and increased open space for conservation and recreation.

Sustainable urbanism is an approach to development that includes green building, high-performance infrastructure, and walkable, transit-oriented neighborhoods. This approach combines green environmental goals with pedestrian-oriented design and development.[ii]

Traditional neighborhood developments (TNDs) are new towns or neighborhoods that have been developed in accordance with new urbanist principles, often in suburban or greenfield contexts. TNDs feature walkable neigborhoods with narrow, well-defined streets laid out in a grid network; a mix of commercial and residential uses; common public spaces; and place-appropriate architecture.[iii]

Transit-oriented development (TOD) refers to residential and/or commercial development oriented around existing or future public transportation. One of the principal goals of TOD is to reduce the reliance on automobile travel and its associated costs.

Walkable urbanism is a term used to describe development patterns that allow residents to walk or use public transport for the majority of their trips; it stands in contrast to *drivable suburbanism*. Both forms of development can exist within or outside cities.[iv]

i. Craig Q. Tuttle, Jill C. Enz, and Steven I. Apfelbaum, "Cost Savings in Ecologically Designed Conservation Developments," Applied Ecological Services, Inc., 2007.
ii. Douglas Farr, *Sustainable Urbanism* (Wiley, 2007).
iii. Robert Steuteville and Philip Langdon, *New Urbanism: Comprehensive Report & Best Practices Guide* (Ithaca, N.Y.: New Urban Publications, 2003).
iv. Christopher Leinberger, *The Option of Urbanism* (Washington, D.C.: Island Press, 2008).

Although approaches to green community development have emerged from numerous directions over the past few decades, they can be categorized as having three main threads: smart growth, new urbanism, and green building. Across the country, government and nonprofit efforts to recognize, encourage, and accelerate the adoption of smart growth have proliferated. These efforts have taken various forms, including regional plans, urban growth boundaries, regional smart-growth alliances, and policies to focus development.[9] Nevertheless, national, state, and local transportation funding is still weighted overwhelmingly toward subsidizing sprawl—in particular, roads and highways. Only 15% to 20% of federal transportation spending is typically allocated to non-automobile forms of transportation, including trains, buses, cycling, and walking.[10]

Led by the Congress for New Urbanism (CNU), a growing movement of architects, developers, and planners has become actively engaged in designing and building walkable communities, many of which are based on the design of prewar neighborhoods. Roughly 900 of these traditional neighborhood developments (TNDs) have been built or planned in the United States in the last half-century, most since the early 1990s.[11] TND principles underlie a range of projects, from high-end resort communities to plans for rebuilding areas devastated by Hurricane Katrina. Often, TND initiatives are designed to spur urban renewal and create desirable and walkable downtowns.[12]

Other innovators, working outside the new urbanist approach, have developed high-performance approaches to neighborhood design; one example is conservation development, which offer benefits by reducing infrastructure needs, protecting watersheds and habitats, and improving property values and quality of life.

In 2007, in response to the large demand for environment- and people-friendly community design approaches, the USGBC, the CNU, and the NRDC launched a pilot LEED for Neighborhood Development program (LEED-ND). LEED-ND guides and certifies developments, awarding credits for both green building and green neighborhood design features. Like other LEED rating systems, LEED-ND provides an independent verification of the greenness of design features—and a label that, it is hoped, will be trusted and valued in the marketplace.[13] Figure 2.2 shows the relative number of points awarded for each category in the LEED-ND system. As can be seen in the figure, the focus is on both *how* and *where* development occurs. LEED for New Construction, the most commonly used LEED standard for individual green buildings, awards one point for locating in a dense mixed-use area. Only 20% of green buildings in the study data set earned that point, reflecting the fact that most green buildings are not built in green neighborhoods, and underscoring the need for LEED-ND's focus on location and community design.

The great diversity of green development approaches is evident in the pool of applicants for the LEED-ND pilot program, which included projects encompassing less than an acre and projects encompassing several thousand acres.[14] Sixteen developments that had

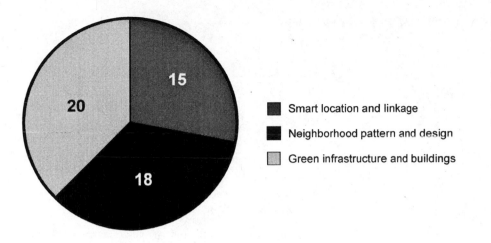

Smart location and linkage

Neighborhood pattern and design

Green infrastructure and buildings

applied to participate in the LEED-ND pilot program supplied basic project information for this study. These projects ranged from a greenfield resort and residential communities to infill redevelopment projects in urban settings. Residential densities varied greatly—from 6 residential units per acre to over 100 units per acre—but all had densities at least double the 3 units per acre typical of conventional developments.

Section 2.2 traces the history of the three main threads of green community design.[15] Its author, Douglas Farr—a green architect, planner, and co-chair of LEED-ND—coined the term *sustainable urbanism,* a design philosophy that yields green communities and unites smart growth, new urbanist, and green building approaches.

2.2. Setting the Stage for Sustainable Urbanism

Douglas Farr

Sustainable urbanism grows out of three late-20th-century reform movements that emphasize the benefits of integrating human and natural systems. The smart growth, new urbanism, and green building movements provide the philosophical and practical bones of sustainable urbanism. While all three share an interest in economic, social, and environmental reform, they differ greatly in their history, constituencies, approach, and focus.

Each of these movements, highly worthy in and of itself, has suffered from a certain insularity, which has resulted in myopia when it comes to searching for long-term, comprehensive solutions. For instance, a certified green building isn't really a positive for the environment when it is surrounded by a massive paved parking lot; a walkable neighborhood is hard to sustain when its houses are wastefully constructed and energy-inefficient. Sustainable urbanism attempts to bring these three important movements together and knit them into a design philosophy capable of fostering and creating truly sustainable human environments.

SMART GROWTH: THE ENVIRONMENTAL CONSCIENCE OF SUSTAINABLE URBANISM

Smart growth has its roots in the environmental movement of the 1970s, which was strengthened by President Richard Nixon's environmentally focused legislative agenda. With bipartisan support, Nixon signed into law what serves as the backbone of United States environmental policy to this day. The list of environmental initiatives includes the creation of the EPA, and the passage of the Clean Water Act, the Clean Air Act, the Endangered Species Act, the National Environmental Protection Act (NEPA), and the Coastal Zone Management Act. In 1970, amid this unique burst of federal environmentalism, Senator Henry (Scoop) Jackson introduced the National Land Use Policy Act.[16] Designed as a bookend to NEPA, it was intended to encourage states to develop coordinated state land use plans; it also proposed the creation of a new federal agency and land-planning database. Although the act failed to pass, several pioneering governors adopted the idea of statewide land use planning.

Building on Oregon's long tradition of land conservation and on its interest in preserving the state's scenic beauty, Governor Tom McCall proposed legislation to manage the state's population growth and land development. In 1973, the Oregon legislature passed a law requiring all municipalities to designate urban growth boundaries—rings beyond which land development was not permitted.[17] Urban growth boundaries succeeded in controlling the scope of land development, thus preserving the state's scenic treasures, but did little to ensure the quality of development within the boundary; the result was well-located bad development, or what could be called "smartsprawl."

Other states took different approaches to regulating land use. Colorado governor Roy Romer first used the term *smart growth* in 1995, in putting forward a vision for an alternative to sprawl. Under the leadership of Governor Parris Glendening, the state of Maryland picked up the term, and in 1997 enacted the Smart Growth and Neighborhood Conservation Program, which designated urban growth areas that were eligible for state infrastructure funding. While the law remained in effect only until shortly after Governor Glendening stepped down, in January 2003, this strategy encouraged other states—notably New Jersey—to follow suit. Maryland's development location criteria would later help to inform similar criteria in LEED-ND.

In 1996, with the development of the ten principles of smart growth, the smart growth movement embraced a broader agenda—an effort that was initiated by Harriet Tregoning, then director of the Development, Community, and Environment Division at the EPA. The principles successfully united a decentralized grassroots movement of local and regional citizen activists and municipal leaders under the smart growth banner; but the vagueness of the standards and the smart growth movement's decision to lend its name to development projects that sometimes yielded minimal incremental improvements worked to devalue the smart growth "brand." Nonetheless, this national coalition of regional, not-for-profit organizations has a dedicated membership promoting urban redevelopment and sound land conservation policies.

CONGRESS FOR THE NEW URBANISM: SUSTAINABILITY'S URBAN DESIGN MOVEMENT

The Congress for the New Urbanism was founded by six architects—Peter Calthorpe, Andrés Duany, Elizabeth Moule, Elizabeth Plater-Zyberk, Stephanos Polyzoides, and Daniel Solomon—and first met as an organization in 1993. The founding members united around a shared vision of promoting traditional urbanism as an antidote to conventional sprawl and created an ad hoc organization to convene four annual congresses.

To best understand CNU, it helps to go back to 1928, and the founding of the Congrès Internationale d'Architecture Moderne (CIAM), or the International Congress of

Modern Architecture. Like CNU, CIAM was a design reform movement with a stated focus of bettering public health and design by improving cities and housing. At its core, the CIAM movement was a humane and essential attempt to improve human health and sanitation; at the time, large sections of the older cities of Europe were dangerous and unhealthy places to live, especially for the lower classes. CIAM's analysis accurately captured the gravity of the problem, citing "a mortality rate reaching as high as twenty percent" in some city quarters.[18]

CIAM was the source for the "towers in the park" pattern of public housing development, which was widely followed in the United States in the years after World War II, and which proved to be isolating for residents; outside of New York City, such housing developments have largely been dismantled. A particularly notable aspect of the towers-in-the-park approach to design is the central place assigned to the needs—one might say the rights—of drivers to travel at high speeds, unimpeded by constraints. At the expense of pedestrians and a fine-grained street grid, this design approach took the poor acceleration and braking of early cars and transformed it into a fundamental basis for street design.[19]

The CNU founders found direct ties between CIAM's vision of a so-called rational city and the postwar American suburbs' automobile dependence and segregated land uses. Streets designed for high speed, the segregation of land uses, and stand-alone buildings were all required in the standard municipal regulations that still shape the sprawl of today. Furthermore, the ascendancy of modernist architectural training essentially erased all knowledge of pre-CIAM town planning techniques.[20] So when the CNU began to promote traditional town planning as an alternative to sprawl, it was largely forced to start from scratch.

A founding goal of the CNU was to write a charter that would rebut CIAM and serve as the governing document for a new reform movement. The CNU principles were later adopted as the centerpiece for the HOPE VI program of public housing revitalization—which aimed, fittingly, to dismantle and rebuild CIAM-inspired postwar public housing developments. This robust housing and community rebuilding program proved vital, introducing new urbanist principles to the real estate industry nationally and creating a market for new urbanist development.

Throughout the 1990s, new urbanism became an increasingly large part of mainstream development practice, in part through the Urban Land Institute's repackaging of new urbanist work as "master-planned communities" or "lifestyle centers." CNU has excelled at creating mixed-use neighborhood developments and transit-oriented villages that feature town centers; fine-grained, walkable street grids; and a diverse ensemble of traditional buildings and architectural styles. Because the projects are routinely deemed illegal under local zoning laws and go against most conventional development practices, the new urbanists have pioneered new approval techniques (notably the town planning charrette).[21]

The desire to control the long-term placement and design of buildings led to the development of form-based zoning, a coherent, green, and pedestrian-friendly approach to zoning.

The new urbanism has also developed significant new approaches to, and tools for, regional planning, a particularly challenging area owing to the lack of regional government and planning authority in the United States. One example is the Envision Utah process, which Peter Calthorpe's firm has successfully used to plan a large number of major metropolitan regions. The most successful of these plans have proven effective at influencing large regional investment decisions, such as transit system funding, road and highway alignments, and overall land use development patterns.

Two other new urbanist innovations, the urban-rural transect and the Smart Code, both developed by Andrés Duany, principal of Duany Plater-Zyberk, also have the capacity to guide design at the regional level.[22] The urban-rural transect applies an ecological framework to describe human settlements along a spectrum ranging from wilderness to dense urban centers. The Smart Code, a transect-based, form-based code, combines aspects of conventional zoning codes, subdivision codes, and overlay districts into one integrated document; the intent is to replace existing zoning codes with clear, simple, and coherent new codes. Within just a few years of its development, numerous local governments had adopted the Smart Code as the basis for their land development controls.

Despite its many achievements, the CNU has proved only moderately successful in reforming state or national practices, in large part because of a focus on convincing local regulators to create exceptions to conventional practice to allow the approval of individual projects. While effective on a case-by-case basis, this pragmatic approach has left intact zoning regulations and standards that are hostile to green design and conservation developments, as well as a built environment that remains largely dominated by climate-changing sprawl.

THE U.S. GREEN BUILDING COUNCIL: SUSTAINABILITY'S BUILDING PERFORMANCE AND CERTIFICATION MOVEMENT

The oil shocks of the 1970s jump-started a movement for energy efficiency and the use of solar heat and power. Throughout the 1980s, such efforts were unable to attract much governmental policy support, and gained little traction. In 1993, however, inspired by the 1992 Rio Earth Summit, the American Institute of Architects' Committee on the Environment published the *Environmental Resource Guide*. This comprehensive catalogue on the theory, practice, and technology of "environmental" buildings drew heavily on the pioneering work that preceded it. This same confluence inspired the creation of the third founding reform of sustainable urbanism, the U.S. Green Building Council. The USGBC was founded in Washington, D.C., in 1993 by three development industry professionals: David Gottfried, Richard Fedrizzi, and Michael Italiano.[23] They, too, were inspired by the

Rio Earth Summit and were largely concerned with the same intellectual ground explored in the *Environmental Resource Guide*. The USGBC made two very smart moves to accelerate the adoption of environmental (or "green") building practices: it expanded its audience outside the architecture profession, and it sought to mobilize the private sector.

Shortly after its founding, the USGBC drafted pioneering standards for green building, completing a "final" version in 1995. The name Leadership in Energy and Environmental Design (LEED) was adopted for the standards in 1996.[24] The USGBC launched the pilot version in 1998 and its rating system in 2000. The LEED standards combine prerequisites with optional credits that earn points toward an overall score. A helpful early breakthrough was the decision by the U.S. General Services Administration to adopt LEED standards as a requirement for all government-owned and -developed buildings.

The backbone of the success of LEED has been the USGBC's ability to increase its staff and certification operations at a geometric pace while maintaining quality and integrity. The USGBC has successfully mobilized and harnessed a huge amount of volunteer effort from hundreds of professionals, and has trained tens of thousands in LEED standards and practice. So far, LEED has found a middle ground, staving off arguments that LEED documentation was too rigorous on the one hand and no longer cutting edge on the other. Another key aspect of green building practice is integrated design: working in interdisciplinary teams to optimize overall building performance without adding construction cost.

The launch of the LEED-ND pilot program, in the spring of 2007, brought the energy and momentum of the USGBC's green building certifications to the area of community design. Innovations new to LEED-ND include increasing the number of available credits to 100 and adding criteria that award points for walkable streets, affordable housing, and local food production, among other attributes. The market demand for the product was overwhelming: 371 expressions of interest, with a total of 231 domestic (42 states and the District of Columbia) and international (including Canada and China) projects accepted into the pilot. As it was designed to do, the pilot process—by testing the standards on real-world projects—generated a lot of suggested revisions to the rating system. The changes included the addition of many new prerequisites, and the consolidation and reweighting of the credits. LEED-ND 2009 was adopted in 2009, and will participate in biannual updates from then on.

2.3. Financial Impacts of Green Community Design

Green community design has financial consequences for developers, homeowners, and local governments. Given the diversity of approaches and the fact that most green communities are quite young, assessing these impacts requires drawing on a range of research, case studies, and surveys.

FIRST COSTS

As with green buildings, the general perception that green communities require increased up-front costs bears closer examination. A direct comparison of costs is difficult, however, because conventional and green developments differ on many dimensions. For instance, green neighborhood design may increase the number of residential units in a given parcel by 10% to 100% when compared with conventional development, while decreasing lot size, changing the layout of roads and intersections, and adding public spaces, walking paths, and a mix of commercial uses that would typically be excluded from conventional suburban residential development.[25] The perception that green community developments require higher first costs comes from two sources: the more complex nature of mixed-use projects, and the use of green building strategies. For example, in urban infill locations and sites on the edge of existing urban developments—which are often sites for green communities—land costs are (1) typically significantly higher and (2) make up a higher portion of development costs than in the exurban areas where most conventional development occurs.[26]

On the other hand, data from some types of green development show significant first-cost savings on a per-project and per-unit basis. Even when green vertical development (i.e., green buildings) is somewhat more expensive, horizontal development costs can be significantly lower, thanks to features such as narrower streets and natural strategies for dealing with storm water. Analysis of ten conservation developments, in "Cost Savings in Ecologically Designed Conservation Developments" (section 2.8), demonstrates substantial first-cost savings.

An informal survey of LEED-ND pilot applicants, conducted in 2006 and 2007 for this book, found *anticipated* additional costs of 2% to 20% of total project costs. Surveys of private and public leaders in green design, construction, and development show similar general expectations. A 2000 survey of 23 industry leaders, for example, indicated a wide-spread perception that mixed-use TND projects have higher initial costs than conventional single-use residential or commercial development.[27] Increased development costs were generally thought to be up to 10%, and to result from higher-quality design, increased density, multiple uses, and nonstandard designs. Investors and lenders participating in the survey did not generally perceive the increased cost to be a major barrier to "the financing of well-conceived" TNDs.[28]

Perceived risk is equally important for the financing of green communities. The greater complexity of mixed-use projects is widely assumed to increase risk. Surveys of real estate experts indicate that more complex TND projects require greater management skill and can increase the complexity of financing and project management.[29] TNDs that are not adjacent to existing towns can take many years to reach full build-out, creating additional risk during the period before a critical mass of residential and commercial development creates the pedestrian traffic that is required to sustain local businesses, and before residents are able to meet a variety of their needs locally (restaurants, shopping, schools, jobs, etc.).

Mixed-use infill developments, although distinct from TNDs in form, face similar challenges because of the diversity of uses and the need for nonstandard design. In 2006, as part of the Conference on Mixed-Use Development, 1,000 members of development organizations were surveyed about mixed-use development.[30] Seventy percent of respondents believed that greater risk is associated with mixed-use development, and over 60% believed that mixed-use projects cost more than an equivalent number of separate components. "Assembling land and parcels," "maneuvering through zoning regulations," and "managing the financial challenges of a sequenced roll-out of project parts" were rated as the top three challenges to mixed-use developments.[31]

Many of the increased costs associated with green communities result from zoning codes that prohibit or discourage mixed-use development, or that require or show a preference for conventional sprawl. As a result of these obstacles and the risk inherent in them, developers see greater challenges in financing mixed-use projects, though a large majority of those surveyed see lenders as becoming increasingly knowledgeable and comfortable financing mixed-use projects.

The complexity of first development costs in green versus conventional developments is surely not fully captured by the limited research and surveys conducted to date. As noted earlier, analyses developed for this book, which cover ten recent ecological conservation developments comprising over 1,500 homes, demonstrate average savings of $12,000 per home compared with conventional sprawl.[32] First-cost savings associated with conservation developments derive largely from infrastructure and can offset other design features such as enhanced public spaces, permeable pavement, district-based renewable-energy systems, and rainwater harvesting. For instance, similar infrastructure savings are beginning to be documented in "light-imprint" approaches to TND—dense, mixed-use communities that use some of the infrastructure techniques learned from conservation development.[33]

INFRASTRUCTURE COSTS

Compared with conventional sprawl, green communities present significant opportunities to reduce initial and long-term infrastructure costs for both developers and local governments. In green developments, clustered buildings and smaller lot sizes typically reduce the need for new sewer, water, and utility connections; reduce land-clearing and grading costs; and create long-term public savings from the reduced need to maintain infrastructure and manage storm water. A potentially higher cost for roads in some dense, grid-connected neighborhoods is usually more than offset by savings on other infrastructure.[34] In many green communities, pedestrian walkways and bike paths can replace some streets, further reducing infrastructure costs. To the extent that infrastructure costs are borne by developers, lowering such costs may offset some or all of the additional costs of green development.

The value of up-front infrastructure savings varies, depending on the characteristics of the site and the project. An analysis of three alternative approaches to coastal development in South Carolina, conducted by the National Oceanic and Atmospheric Association (NOAA), found that a TND approach reduced up-front infrastructure costs by 9% on a project basis and 7% on a per-unit basis (see figure 2.3).[35]

FIGURE 2.3 Development Cost Comparison: Alternatives for Coastal Development
Source: National Oceanic and Atmospheric Administration (NOAA), "Alternatives for Coastal Development: One Site, Three Scenarios," NOAA, 2005 (www.csc.noaa.gov/alternatives/).

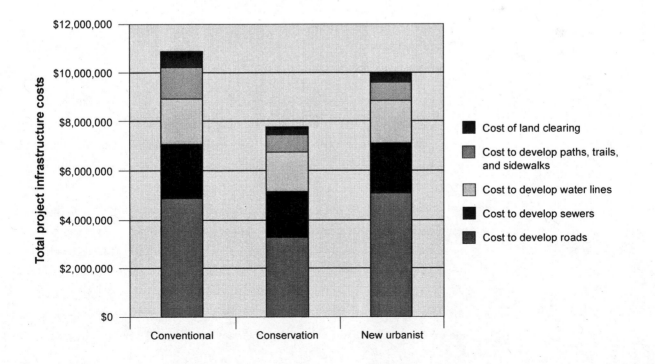

Developing in infill locations—a green choice rewarded by LEED-ND requirements, provides very substantial infrastructure savings. Studies by the EPA and others have found that per-unit infrastructure costs for infill development can be 90% lower than for greenfield development—which translates to as much as $25,000 per residence.[36] Put another way, per-unit infrastructure costs may be up to ten times greater in car-dominated suburban sprawl than in infill development settings—largely because entire categories of infrastructure that are needed for greenfield development, including roads, and sewer and water systems, are already present at infill locations. The infrastructure savings associated with infill development are particularly important because long-term infrastructure costs are almost always borne largely by the community, rather than by developers.

A 1997 study by the Canada Mortgage and Housing Corporation, which examined life-cycle costs over 75 years for TNDs and conventional sprawl, found that with TNDs, public sector costs were 48% lower for commercial buildings and 5% lower for residential uses.[37] Another analysis, undertaken in Salt Lake City, compared sprawl with a slightly more compact scenario for future development, and estimated that over 20 years, the potential reduction in infrastructure costs for the compact development scenario was 25%, or $4.5 billion.[38]

Additional community-wide benefits result where green developments create public amenities such as parks, trails, and protected habitat. Green communities typically have substantially more green space than conventional developments, and may include features such as nature centers, community gardens, or farmers' markets. In contrast to sprawl, which places unsustainable infrastructure burdens on local governments, green community design can create additional spaces for public recreation at no cost to the jurisdiction.

STORM WATER

TNDs and other green development patterns significantly reduce storm-water flow per home, primarily through increased density. A 2006 EPA study found that doubling residential density decreased storm-water runoff by up to 67% *per unit*, by reducing total impermeable area and leaving more land available for rainwater infiltration.[39] In addition to decreasing infrastructure costs, reduced storm-water flow provides additional benefits: reduced source pollution, reduced flooding, and related reductions in downstream impacts on natural systems and human development. Costs for storm-water conveyance and treatment are highly site specific; as a consequence, so are the benefits of reducing storm-water flow. A case study in South Carolina by NOAA, the EPA, and Dover, Kohl and Partners compared conventional and TND plans for a single site and found that the TND plan, even though it accommodated a greater number of homes, resulted in a 30% reduction in runoff, a 73% reduction in sediment leaving the site, a 69% reduction in nitrogen, and a 67% reduction in the phosphorus pollution present in the runoff.[40] The TND scenario allowed much more land to remain undeveloped by decreasing road widths and clustering development, which reduced storm-water flow while providing benefits in terms of transportation, walkability, and access to nature.

Green techniques, including bioswales, pervious pavement, habitat restoration, and biological storm-water treatment have the potential to further reduce storm-water flow and related pollutants. High Point, a HOPE VI urban redevelopment project in Seattle, incorporates bioswales, pervious pavement, and narrow streets, reducing storm-water runoff into nearby Longfellow Creek by 65%.[41]

Another case study of a planned subdivision, Remlick Hall Farm, outside Washington, D.C., found that clustered development led to a 53% reduction in road length, and a $525,000 savings in development costs for the 84-residential-lot project.[42] These results are generally in line with savings found in the ten projects discussed in section 2.8, "Cost Savings in Ecologically Designed Conservation Developments." Conservation developments typically reduce the annual amount of water leaving the landscape by 50% to 80%, even when compared with conventional developments that have current best practices for storm-water management in place.

Estimating the financial value of storm-water reduction in green communities involves considering (1) reduced costs for conventional drains, basins, and other conventional storm-water infrastructure; and (2) the reduced downstream effects of flooding, pollution, and erosion.[43] Some building codes require unnecessary conventional storm-water infrastructure to be installed even when alternative strategies are in place, preventing some of the potential cost savings associated with green design.

Reduced storm-water flow can also increase general property values in a watershed by reducing the risks and costs of storm-water runoff. A 2006 study of a watershed near Chicago found that conservation development practices produced economic benefits of $380 to $590 per developed acre, from reduced flooding and drainage-infrastructure costs alone.[44] These benefits accrue in the form of increased downstream property values, and reduced municipal costs for storm-water infrastructure.

Reduced or eliminated pesticide use in green communities is likely to have additional health benefits—especially for children, pets, and the waterways into which pesticide residue can be transported. The value of these benefits is not estimated in this book.[45]

PEDESTRIAN SAFETY

Narrow streets, and streetscapes lined by trees, buildings, and other features result in fewer pedestrian injuries and fatalities than the wide streets common in conventional suburban developments, thus reducing the costs and suffering associated with traffic accidents.

For decades, traditional industry practice and the official design guidelines from the

American Association of State Highway and Transportation Officials (AASHTO) have encouraged widening roadways and shoulders. Recent studies by AASHTO and independent researchers, however, have shown that wide lanes and clear shoulders are associated with increases in average driving speed, and with greater frequency and severity of accidents.[46] For example, AASHTO's "Green Book," the handbook of transportation engineers, indicates a 31-mile-per-hour minimum speed in central business districts and intermediate areas. Yet research has found that as traffic speeds climb from 20 to 30 miles per hour, pedestrian fatalities rise 7.5 times, and that at speeds above approximately 36 miles per hour, a majority of pedestrian accidents are fatal to the pedestrian.[47]

A 2005 study compared crash data from historic and conventionally designed sections of a road near Orlando, Florida.[48] The sections were similar or identical in a number of attributes: speed limits, median road widths, number of lanes, and amount of daily traffic; but the historic section had narrower lanes, with on-street parking and buildings abutting the roadway, while the conventional section had wide shoulders free of obstructions. Over five years of recorded data, the historic section experienced 11% fewer mid-block crashes (73, versus 82 in the conventional section), 31% fewer injurious crashes (42 versus 61), and 100% fewer fatalities (0 fatal crashes, versus 1 in the conventional section). A similar analysis applied to records of hundreds of crashes on four other Florida roadways with comparable conventional and historic sections showed similar results; over a five-year period, researchers found 11% fewer accidents, 24% fewer injuries, and 100% fewer fatalities along narrow, well-defined roadways.[49]

Analysis of 20,000 accident reports from 1989 to 1997 in Longmont, Colorado, found that a two-foot increase in lane width was associated with a 35% to 50% increase in accidents resulting in injuries.[50] The number of accidents resulting in injuries was 485% higher on 36-foot-wide streets than on 24-foot-wide streets.

Car accidents have enormous economic costs. A National Highway Traffic Safety Administration (NHTSA) study of crash data, conducted in 2000, estimated that in that year alone, the economic costs of 41,821 fatalities, 5.3 million nonfatal injuries, and 28 million damaged vehicles were $230.6 billion, including property damage, medical expenses, travel delay, and lost productivity. The costs associated with the 5.3 million injuries make up almost 63% of the total costs—$146 billion dollars, or roughly $28,000 per injury.

Most neighborhoods built before World War II—and new, walkable town and urban centers and TNDs—have narrow, well-defined roadways that are relatively pedestrian friendly. Streets in the residential areas of TNDs are typically no more than 24 feet wide, compared with at least 35 feet in conventional suburban development,[51] a reduction that some studies have associated with an 80% reduction in injurious crashes.

A 2006 survey of U.S. highways by the Federal Highway Administration (FHWA) found that 90 injuries occur on U.S. roads per 100 million vehicle-miles traveled (VMT).[52]

Using this figure as the baseline injury rate and assuming 20,000 VMT per year for the average household, a 50% reduction in crashes would be worth approximately $250 per household per year, taking into account only the direct physical costs of the injury and crash. The discounted present value of this benefit over 20 years would be $3,300 per household, or $1.50/sf, assuming an average household size of 2,200 square feet.[53] The NHTSA study of the costs of crashes emphasizes that this type of cost estimation does not include emotional or societal costs.

As noted earlier, residents of green communities also drive less, resulting in a reduction in VMT. Assuming a 25% reduction in VMT and a 50% reduction in crashes in green communities, the total annual savings from reduced car accidents and injuries would total $300 per household per year, which has a present value of $4,100 per household over 20 years. Furthermore, although some vehicle trips will be replaced by transit or walking trips, these are likely to be considerably shorter, and public transport and walking are considerably safer than driving. FHWA records show that in 2004, 1.44 fatalities occurred per 100 million VMT (in cars), and 0.55 fatalities per 100 million passenger-miles traveled on public transport.[54] Rates of injury are similarly lower when public transport is compared with car travel. Thus, green communities reduce both the risk of fatalities and injuries per mile, and the total number of miles traveled, creating much safer places to live and work.

2.4. Transportation and Health Impacts of Green Community Design

This section explores the transportation- and health-related benefits of green community design. The first subsection, written by leading transportation researchers Lawrence Frank and Sarah Kavage, highlights some of the recent research linking regional location and community design to transportation patterns, CO_2 emissions, non-CO_2 air pollutants, obesity, and physical activity. The second subsection provides preliminary estimates of the value of transportation and health benefits for a two-car household in walkable versus car-dominated neighborhoods in Atlanta.

ESTABLISHING THE LINK BETWEEN TRANSPORTATION, HEALTH, AND GREEN COMMUNITY DESIGN

Green community design makes it convenient to access destinations using a variety of transport modes—walking, biking, transit, *and* driving. In contrast, conventional sprawl effectively makes driving the only practical means of transport. A wide range of studies have documented that green community design is associated with significant decreases in per capita driving and emissions, and with increases in physical activity.[55] A 2007 meta-analysis concluded that compact, walkable development "has the potential to reduce vehicle miles traveled per capita by anywhere from 20% to 40% relative to sprawl."[56] These positive health and environmental outcomes are linked to two aspects of a development or community (note that these correspond to the first two categories of LEED-ND points): regional location and community design.

The location and size of population and employment centers—and the distances between them—shape commuting patterns. Travel times by auto and transit can also affect the relative accessibility of these major origins and destinations. As shown in figure 2.4, community design impacts travel primarily by creating greater proximity between neighborhood destinations and increasing the directness of travel between these destinations.[57] Green neighborhood design typically incorporates a variety of elements that support a range of mobility options, including compact, walkable design; a mixed-use development pattern in which homes, shops, and services are within walking distance; an interconnected street network; and access to public transportation.

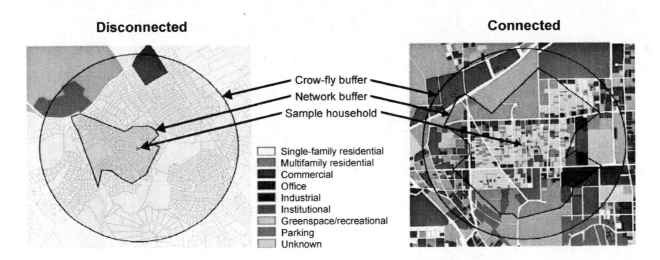

Disconnected   **Connected**

Crow-fly buffer
Network buffer
Sample household

- Single-family residential
- Multifamily residential
- Commercial
- Office
- Industrial
- Institutional
- Greenspace/recreational
- Parking
- Unknown

FIGURE 2.4 Primary Neighborhood Design Characteristics Linked to Travel Behavior
On the left, a household located in a typical single-use, low-density suburban neighborhood with a disconnected street pattern; on the right, a household located in a compact, connected, mixed-use neighborhood. The circle represents a one-kilometer radius (the "crow-fly" distance) from each household; the asymmetrical "network" buffer inside the circle captures the one-kilometer area that is actually walkable on the street network. The figure illustrates not only how a disconnected street network pattern limits the area that is accessible on foot (directness), but also how a low-density, single-use land use pattern restricts the number of accessible destinations within walking distance (proximity).
Source: Lawrence Frank, Martin Andersen, and Thomas L. Schmid, "Obesity Relationships with Community Design, Physical Activity, and Time Spent in Cars," *American Journal of Preventative Medicine* 27, no. 2 (2004): 87–96.

The remainder of this section is structured as follows: we ask a series of questions about the various impacts of green communities, then highlight recent research that addresses each question, using two overall topic headings: regional location and community design.

HOW MUCH DO GREEN COMMUNITIES REDUCE VEHICLE USE?

Regional Location

A national analysis found that the degree of sprawl in a region is the single strongest influence on per capita VMT—stronger, for example, than metropolitan population or per capita income.[58] In a study of Chicago, Los Angeles, and San Francisco land use and travel patterns, researchers found that regions with twice the density have 25% to 30% less driving per household, provided that conditions associated with density—such as transit, nearby shopping, and pedestrian amenities—are present.[59]

The map of Atlanta shown in figure 2.5 clearly reveals these regional patterns. The farther people live from the region's central cities, the higher the VMT.

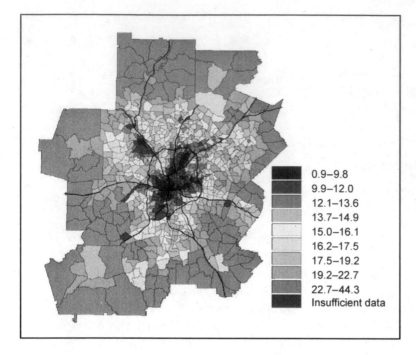

	0.9–9.8
	9.9–12.0
	12.1–13.6
	13.7–14.9
	15.0–16.1
	16.2–17.5
	17.5–19.2
	19.2–22.7
	22.7–44.3
	Insufficient data

FIGURE 2.5 Daily Per Capita Home-Based Vehicle-Miles Traveled: Atlanta Region, 1998
Source: Georgia Regional Transportation Authority, 1998.

Community Design

Six elements of green community design—density, land use mix, street connectivity, walkability, site design, and transit service—have strong links to reductions in VMT, hours spent driving, likelihood of pedestrian travel, and likelihood of transit use. The following list provides a sampling of research findings in these areas.

- **Density.** All other things being equal, as density increases, the number of vehicle trips and the per capita hours and miles of automobile travel decrease. In a study of three cities (Chicago, Los Angeles, and San Francisco), J. Holtzclaw et al. found that each doubling of residential density reduced VMT by 32% in Chicago, 35% in Los Angeles, and 43% in San Francisco.[60]

- **Land use mix.** The Seattle-area Land Use, Transportation, Air Quality, and Health (LUTAQH) study found that increasing the mix of land uses in an area from the lowest to the highest quartile was associated with a 19.7% decrease in VMT and a 23.5% decrease in vehicle-hours traveled.[61]

- **Connectivity.** The LUTAQH study also found that each quartile increase in connectivity, as measured by the number of intersections per square kilometer, corresponded to a 14% increase in the odds of walking for nonwork travel.[62] When controlling for demographic variables, mean daily VMT was 34 miles per person in the least connected environments, and 25 miles per person in the most connected environments—in other words, for residents living in communities with the most interconnected street and walking networks in the county, VMT was 26% lower.[63]

- **Walkability.** By integrating several different measures of urban form into one index value, it is possible to account for covariation in the different factors that affect urban form, and also for any synergistic effects that may occur when all characteristics appear together, as they often do. The Atlanta-based SMARTRAQ (Strategies for Metropolitan Atlanta's Regional Transportation and Air Quality) study used a walkability index that included residential density, street connectivity, and land use mix. People who lived in neighborhoods that scored lowest on this index drove an average of 39 miles per person each weekday—30% more than those who lived in the most walkable neighborhoods.[64]

- **Site design and the pedestrian environment.**[65] If a building is placed adjacent to the sidewalk, walking distances are shorter and the walking environment is more pleasant than if there is a parking lot between the building and the sidewalk. The presence of sidewalks, street trees, benches, and other features can increase pedestrian safety, convenience, and comfort and provide an attractive environment for walking. The Land Use, Transportation, and Air Quality study undertaken in Portland, Oregon, found that "a 10% reduction in vehicle-miles traveled can be achieved with a region-wide increase in the quality of the pedestrian environment"—an environment that is comparable to Portland's most pedestrian-friendly areas.[66]

- **Transit service.** The LUTAQH study found that for every quarter-mile increase in distance from one's *home* to a transit stop, the odds of taking a transit trip to work decreased by 16%. The impact of distance from one's *work* to transit was twice that, with every quarter-mile increase in distance reducing the likelihood of taking transit to work by 32%.[67] A different study in the same region found that reducing travel time was the most important variable in inducing a shift from driving to public transit and walking. For example, faster and more frequent transit service was associated with higher rates of transit use, while shortening highway travel times *for cars* was associated with lower proportions of trips made on foot and by transit.[68]

HOW MUCH ARE CO_2 EMISSIONS REDUCED IN GREEN COMMUNITIES?

Researchers have closely studied the relationships between land use patterns, air pollution, and, to some extent, energy use. Recently, a few studies have begun to look directly at the connection between land use and GHG emissions, although few have taken a detailed look at the specific land use characteristics that have an impact on CO_2.

Regional Location

One researcher who used a model to estimate the effect of land use policies on GHG emissions suggests that smart growth policies could cut transportation emissions by 35% by 2050.[69] As in the map of the Atlanta region shown in figure 2.5, per capita CO_2 generation also increases the farther people live from central cities. In Chicago's case, pockets of lower per capita CO_2 generation can be seen around the regional rail lines (see figure 2.6).

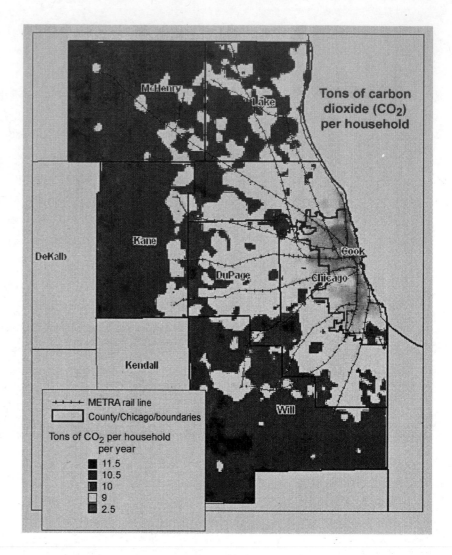

Community Design

Elements of community design that are associated with reduced driving are also associated with reduced transport-related CO_2 emissions. Figures 2.7 through 2.10, which are based on data from King County, Washington, show the relationship between CO_2 emissions and retail availability, land use mix, street connectivity, and density. The analyses controlled for age, gender, education, income, and percentage of population with drivers' licenses.[70] The photos illustrate the types of development associated with various community attributes.

As driving decreases, per capita emissions of other, non-CO_2 air pollutants—such as nitrogen oxides (NOx), carbon monoxide, VOCs, and particulates—also decrease. For instance, each step up the walkability index developed for the Atlanta region was associated with a 6% reduction in NOx and a 3.7% reduction in VOCs.[71] Reductions in air pollutants are associated with lower incidence of respiratory illness and asthma.

FIGURE 2.7 Retail Availability and Mean Daily Transport-Related Carbon Dioxide Per Person: King County, Washington

Source: Lawrence Frank, "Planning for Climate Change" (presentation, 2008 American Planning Association Conference, Las Vegas, Nev., April 28–May 1, 2008).

Note: Analysis controlled for gender, age, education, income, and drivers' license availability.

FIGURE 2.8 Land Use Mix and Mean Daily Transport-Related Carbon Dioxide Per Person: King County, Washington

Source: Lawrence Frank, "Planning for Climate Change" (presentation, 2008 American Planning Association Conference, Las Vegas, Nev., April 28–May 1, 2008).

Note: Analysis controlled for gender, age, education, income, and drivers' license availability.

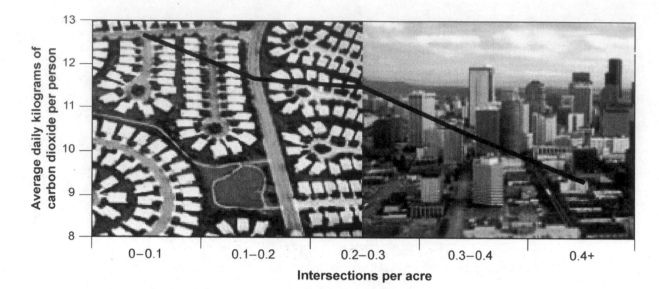

FIGURE 2.9 Street Connectivity and Mean Daily Transport-Related Carbon Dioxide Per Person: King County, Washington

Source: Lawrence Frank, "Planning for Climate Change" (presentation, 2008 American Planning Association Conference, Las Vegas, Nev., April 28–May 1, 2008).

Note: Analysis controlled for gender, age, education, income, and drivers' license availability.

FIGURE 2.10 Net Residential Density and Mean Daily Transport-Related Carbon Dioxide Per Person: King County, Washington

Source: Lawrence Frank, "Planning for Climate Change" (presentation, 2008 American Planning Association Conference, Las Vegas, Nev., April 28–May 1, 2008).

WHAT IS THE IMPACT OF GREEN COMMUNITIES ON RATES OF OBESITY AND PHYSICAL ACTIVITY?

Sprawl has been correlated with higher body weight, obesity, and associated increases in diabetes and cardiovascular disease, among other health problems.[72] Studies have found an association between urban form and the amount of active transport—walking and biking—that occurs, and with the total amount of physical activity.[73] Neighborhoods that generate the most walking trips are ones where daily activities (home, work, school) are located near those that are used less regularly (movie theaters, shops, restaurants).[74]

Regional Location

A 2004 study of 100 metro areas across the United States correlated the level of sprawl with 16 different chronic diseases, including abdominal problems, severe headaches, conditions that have been linked to excessive weight (e.g., hypertension), and respiratory ailments (e.g., emphysema and asthma). Higher levels of sprawl were found to be significantly associated with an increased number of chronic medical conditions.[75]

Community Design

Four elements of green community design—density and land use mix, transit service, walkability, and recreational facilities and open space—have strong links to increased physical activity and lower levels of obesity. The following list provides a sampling of research findings in these areas.

- **Density and land use mix.** The Atlanta SMARTRAQ study found that greater density and a greater mix of land uses were significantly associated with less driving and more walking. Each additional hour a day spent in a car was also associated with a 6% increase in the odds of being obese, and each additional kilometer walked a day was associated with a 4.8% reduction in the odds of being obese. The study showed that an average (5'10") white male living in the areas of the Atlanta region with the least mix of uses weighs approximately ten pounds more than his counterpart in the areas with the highest mix of uses in the region.[76]

- **Transit service.** Research indicates that good public transit service can encourage physical activity, in part because most public transit trips also involve a walking link. One analysis of U.S. travel survey data found that 16% of all recorded walking trips were part of transit trips, and that transit-based walking trips tended to be longer than average.[77] Another study found that 29% of U.S. transit users walked more than 30 minutes daily on their transit trips alone.[78]

- **Overall neighborhood walkability.** In the Atlanta-area SMARTRAQ study, the walkability index was a significant factor in explaining the number of minutes per day of moderate physical activity. Residents of the most walkable areas of the Atlanta region were found to be 2.4 times more likely to have the recommended 30 minutes of moderate physical activity per day.

- **Recreational facilities and open space.** Research indicates that residents of communities with parks, trails, playing fields, and other recreational facilities within walking distance are more likely to be physically active. One study showed that respondents were more likely to achieve the recommended amount of daily physical activity if they lived within a ten-minute walk of a park, trail, or other place to walk.[79] Another survey found that trail use decreased by almost 50% with every quarter-mile increase in access distance.[80]

VALUING THE TRANSPORTATION AND HEALTH IMPACTS OF GREEN COMMUNITIES

This section offers preliminary estimates of the value of transportation and health benefits for a two-car household in walkable versus car-dominated neighborhoods in Atlanta. Estimated net savings are on the order of $800 per year in reduced gas costs per household, and roughly $1,700 per year assuming that 25% of households can eliminate the ownership of one car. For households in the most walkable neighborhoods, health care costs are likely to be reduced an average of $300 per year.[81]

Gas and Car Ownership

The transportation and health impacts of development patterns have significant financial implications. People living and working in green communities can spend less money on gasoline, and may be able to reduce household car ownership. The total savings will vary with the characteristics of the particular community and household. Based on the SMARTRAQ study of the Atlanta region, we estimated that an average two-car household in a highly walkable neighborhood would use 25% less gasoline annually than a household in one of the least walkable neighborhoods. At a cost of $3/gallon, this is a savings of $786 per year in gasoline costs (see table 2.3).[82]

Walkable neighborhoods can also allow some households to reduce car ownership, further cutting expenses. Given average costs of car ownership for the region (including the cost of the car itself plus maintenance, insurance, and gas), the average two-car household in the most walkable neighborhoods in the Atlanta region would save about $4,800 per year from the combination of reduced driving and eliminating the need for one car.[83] However, as car use decreases and public transportation use increases in the most walkable neighborhoods, some car trips will be replaced by public transit, increasing household spending on public transit.[84] Total net transportation savings for a household in a

**TABLE 2.3 ANNUAL GASOLINE COSTS IN MOST AND WALKABLE AND ———
LEAST WALKABLE NEIGHBORHOODS IN ATLANTA (PER TWO-CAR HOUSEHOLD)**

	Least walkable	Most walkable	Difference
Gallons used	1,048	786	262
Dollars spent	$3,144	$2,358	$786

Note: Assumes an average gas price of $3/gallon.

walkable neighborhood are therefore roughly $4,400 per year for households able to avoid the ownership of one car and to increase the use of public transportation. Assuming that 25% of households in the most walkable neighborhoods are able to eliminate one car, we can estimate that average household transportation savings (taking account of lower car-ownership costs [$1,100 per year] and lower gas costs, but increased public transit spending) would be approximately $1,700 per year.

Health Care Savings

In green communities, increased physical activity and reduced obesity, diabetes, and other chronic illnesses result in both direct health care savings and significant improvements in quality of life.[85] These savings accrue to residents, health care providers, government agencies, and employers, and provide significant private and public financial benefits. The U.S. Centers for Disease Control and Prevention (CDC), along with a growing number of urban planners and health professionals, are now actively engaged in training local governments to use health-impact assessments to evaluate and influence planning decisions.[86]

One method for estimating the value of health impacts in green communities is to look at (1) the relationship between community design and level of physical activity, and (2) the relationship between physical activity and health care costs. For instance, as noted earlier, 37% of residents in the most walkable neighborhoods in Atlanta get the recommended 30 minutes per day of moderately intense physical activity, compared with only 18% of residents in the least walkable neighborhoods.[87] According to researchers at the CDC, health costs for individuals who get at least 30 minutes of moderate physical activity at least three days per week are 24% lower than health costs for those who do not.[88] Assuming, conservatively, a similar decrease in health costs for Atlanta residents who achieve at least 30 minutes of moderate physical activity per day suggests that a resident of the most walkable neighborhoods of Atlanta would be expected to spend $114 per year less on health care than a resident of the least walkable neighborhoods. Table 2.4 shows the assumptions used to derive this estimate. (Note

TABLE 2.4 ESTIMATED SAVINGS IN HEALTH CARE COSTS IN MOST ————
WALKABLE VERSUS LEAST WALKABLE ATLANTA NEIGHBORHOODS

	Least walkable neighborhoods	Most walkable neighborhoods	
Percentage of population getting the recommended 30 minutes per day of moderate-intensity exercise[1]	18%	37%	
			Annual savings in health care costs in most walkable neighborhoods
Estimated annual health care costs per person[2]	$2,353	$2,239	$114

Notes:
1 Based on Lawrence D. Frank et al., "Linking Objective Physical Activity Data with Objective Measures of Urban Form," *American Journal of Preventive Medicine* 28, no. 2S (2005).
2 Based on Centers for Disease Control estimates of health care costs in relation to physical activity, adjusted to 2007 dollars.

that the estimate does not include the value of increased productivity—that is, the loss of work avoided through a reduction in health problems.)

Health benefits may be even greater—particularly for youth—where households can eliminate the use of a car. A major study published in 2007 compared health and household car ownership among the 3,161 youth (ages 5 to 18) included in the SMARTRAQ study, and found that when compared with youth from households with three or more cars,

- Youth from households with *two* cars were 1.4 times more likely to walk at least once over a two-day period.
- Youth from households with *one* car were 2.6 times more likely to walk at least once over a two-day period, and 2.2 times more likely to walk more than a half-mile per day.
- Youth from households with *no* car were 7.7 times more likely to walk at least once over a two-day period, and 6.8 times more likely to walk more than a half-mile per day.[89]

Using the calculations comparing health costs in the least and most walkable Atlanta neighborhoods, we can estimate the net present value of 20 years of transportation and health impacts for a 2.6-person, 2,200-square-foot household in a green community that achieves a similar increase in walkability. Using a 7% discount rate and assuming inflation at 2% per year, the value of transportation and health benefits over 20 years, including reduced gas costs, reduced car ownership for 25% of households, and health savings, is close to $25,000 per household, or roughly $10/sf.[90]

Rachel Scheu and Kathryn Eggers, *Center for Neighborhood Technology*

In 2002, the Center for Neighborhood Technology (CNT), a national nonprofit organization, completed a green renovation of its 15,000-square-foot office in Chicago, earning LEED Platinum certification. In designing and constructing the building, CNT intended to demonstrate that LEED Platinum could be achieved at a cost that was comparable to conventional rehab. Strategies for achieving energy efficiency included high-quality construction, a tight envelope, high insulation levels, and high-efficiency systems. Two green demonstration projects, an ice-storage cooling system and photovoltaic solar panels, added $137,800 above typical construction costs. Grants reduced the additional cost to $85,000, yielding a green premium of roughly 7% on the total construction cost.

Monitoring Energy Performance

Based on energy modeling undertaken for the LEED certification process, the building was projected to use 47,000 Btu/sf of energy each year, representing a savings of 54% from the ASHRAE 90.1 1999 baseline used by LEED 2.0. Actual energy use in the building's first three years of operation showed a 46% reduction from the ASHRAE baseline (see figure 1). To assist with monitoring efforts, CNT developed a Web-based tool that calculates and displays energy use, carbon emissions, water consumption, and transportation energy intensity (see figure 2). Monitoring building performance helps verify that buildings meet their design targets, and helps quantify impacts on operating costs, emissions, and occupants.

Health and Comfort

In March 2007, the New Buildings Institute administered an occupant-comfort survey that measured satisfaction with air quality, lighting, temperature, and acoustics. Responses were favorable: for 25 out of 26 questions, over 50% of respondents expressed satisfaction (see figure 3). One employee noted that "the indoor air quality in the summer helped with seasonal allergies as compared to when I was outside or at home." Satisfaction was highest with lighting and air quality, and lower for acoustics and temperature. CNT attributes lower satisfaction with acoustics to the open office environment. In 2008, CNT performed additional diagnostics to measure temperature by location, and administered a follow-up survey to identify changes in occupant comfort.

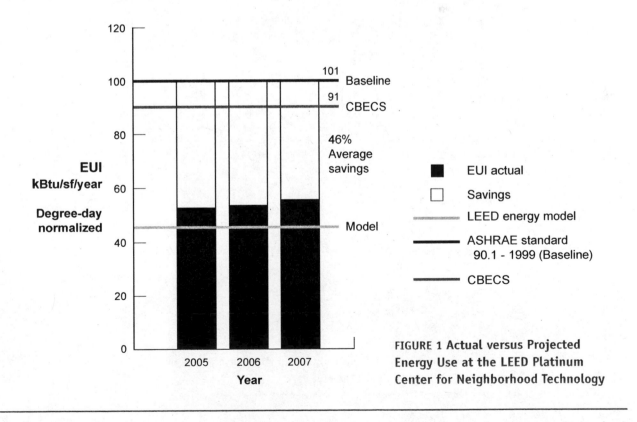

FIGURE 1 Actual versus Projected Energy Use at the LEED Platinum Center for Neighborhood Technology

Transportation Energy Intensity

The energy savings from designing, constructing, and operating high-efficiency buildings may be negated by the energy spent commuting to and from those buildings. To help quantify the energy savings of the decision to stay and renovate, CNT calculates its transportation energy intensity by tracking vehicle-miles traveled for staff commutes. Taking into account staff commuting distances and mode (e.g., car, walking, public transport), CNT's transportation energy intensity is 38,000 Btu/sf per year; its operating energy intensity is 54,000 Btu/sf per year.[1] Because of its proximity to public transportation and staff homes, CNT's transportation energy intensity is 69% lower than the average transportation intensity for U.S. buildings—a greater relative reduction than the 46% reduction in building energy use achieved by the green renovation. The building location plays a significant role in CNT's transportation energy performance.

(continued on page 123)

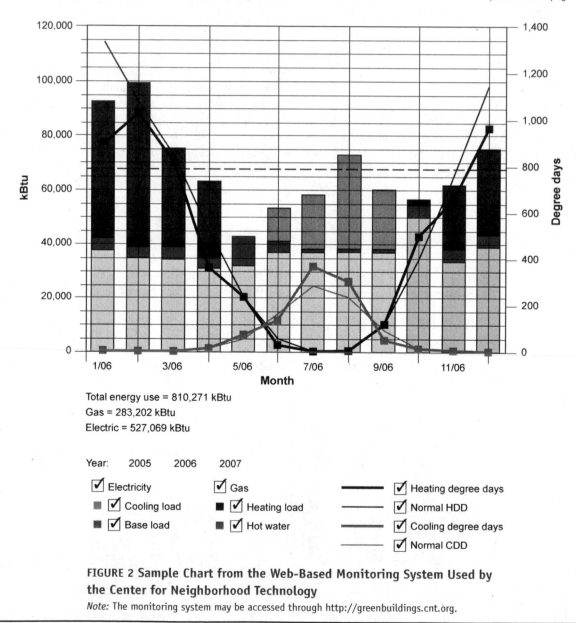

Total energy use = 810,271 kBtu
Gas = 283,202 kBtu
Electric = 527,069 kBtu

Year: 2005   2006   2007

☑ Electricity   ☑ Gas   ▬▬▬ ☑ Heating degree days
▨ ☑ Cooling load   ■ ☑ Heating load   ─── ☑ Normal HDD
▦ ☑ Base load   ▨ ☑ Hot water   ▬▬▬ ☑ Cooling degree days
                                           ┉┉┉ ☑ Normal CDD

FIGURE 2 Sample Chart from the Web-Based Monitoring System Used by the Center for Neighborhood Technology

Note: The monitoring system may be accessed through http://greenbuildings.cnt.org.

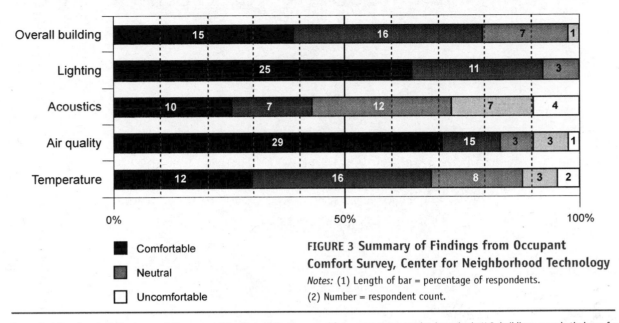

FIGURE 3 Summary of Findings from Occupant Comfort Survey, Center for Neighborhood Technology

Notes: (1) Length of bar = percentage of respondents.

(2) Number = respondent count.

---

1. The methodology for calculating transportation energy intensity, and the figure used for average transportation intensity in U.S. buildings, were both drawn from "Driving to Green Buildings: The Transportation Energy Intensity of Buildings," *Environmental Building News* (September 2007).

Valuing the transportation impacts of community design in this way suggests that even if there is a price premium in choosing to live in a green community, the additional cost is likely to be more than paid back by reduced transportation and health costs over time. This type of trade-off may be especially important for low-income families, for whom transportation costs are a much larger share of income. In 2008, the Center for Neighborhood Technology launched the Housing + Transportation Affordability Index, a mapping tool that gives a fuller picture of affordability for a number of major U.S. cities by taking both housing and transportation costs into account.[91]

2.5. Property Value and Market Impacts

After more than half a century of low-density suburban development as the standard product purchased by homebuyers, recent evidence shows a significant shift in demand toward walkable urban neighborhoods; this shift has been signaled by increased property values in dense, walkable, mixed-use communities. A review of current research suggests that homes in green communities may be worth on the order of 10% more per unit than homes in conventional developments.

CONSUMER PREFERENCE

Recent surveys suggest that while roughly one-third of Americans would prefer to live in walkable neighborhoods, only 5% to 20% of available housing is located in such neighborhoods.[92] For example, a 2005 consumer-preference survey conducted among 1,455 Atlanta homebuyers demonstrated that only 5% of Atlanta residences were located in walkable neighborhoods, up to 30% of respondents expressed a strong preference for living in such neighborhoods.[93] Only 35% of Atlantans surveyed who preferred walkable neighborhoods lived in walkable neighborhoods—suggesting that in Atlanta, demand for housing in walkable neighborhoods was roughly five times greater than availability.[94] A similar survey in Boston showed that 70% of those who preferred walkable neighborhoods actually lived in such environments.[95] While Boston has a greater proportion of walkable neighborhoods than Atlanta, there still appears to be unmet demand for walkable development—even in a place like Boston, a mixed-use city with strong public transport.

Results from a 2008 survey by Robert Charles Lesser & Company (RCLCO) confirm unmet demand for housing in dense, walkable, mixed-use communities:

- 8% of respondents classified their current neighborhood as a TND, but 13% of respondents indicated that they would prefer a TND neighborhood in their next home purchase.
- 68% of respondents classified their current neighborhood as a standard suburban

neighborhood, but only 50% stated that they would prefer a standard suburban neighborhood in their next home purchase.

- 80% of respondents classified their current home as a single-family detached home, but only 68% of respondents stated a preference for a single-family detached home in their next home purchase.

- 59% of respondents—three out of five—preferred multifamily housing to be located in more dense, urban, mixed-use neighborhoods.[96]

PROPERTY VALUE

To the extent that green communities can fill this unmet demand for walkable neighborhoods, they are likely to achieve higher value than conventional, car-dominated neighborhoods. For similar residential properties in walkable urban and car-dominated suburban locations across the United States, a comparison of per-unit and per-acre prices shows substantially greater property values in walkable urban neighborhoods. Indeed, a comparison of per-unit prices within a variety of markets indicates that some consumers are willing to pay premiums of more than 40% for both housing and office space in walkable neighborhoods.[97]

Similarly, TNDs have recently seen increases in house and land value when compared with nearby conventional developments; in some cases, TNDs have experienced increased sales during periods of market-wide decline. An analysis of 2,061 single-family housing transactions that compared prices for homes in Kentlands, a TND in Maryland, with those for a nearby conventional residential development showed a sales premium of 12% for houses in Kentlands.[98] A 1998 review found that revenue from lot sales per acre in Newpoint, a walkable development in South Carolina, was 84% higher than for lots in a nearby comparable conventional subdivision.[99]

Greater property values can also be found among conservation developments—a type of green residential development that features increased open space, ecological restoration, reduced storm-water flow, and lower maintenance costs. When compared with conventional development, conservation development involves significantly smaller private residential lots and significantly larger amounts of shared open space. In the ten conservation developments analyzed for this book (see section 2.8, "Cost Savings in Ecological Conservation Developments"), residential lots are, on average, roughly 60% smaller per unit than those in conventional developments. Yet reports from conservation developers indicate that greater access to open space appears to more than offset smaller lot sizes in determining property values. Orienting homes to open space (e.g., providing more windows, more views, porch to trail access) and access to a range of amenities (e.g., trails, lakes, equestrian activities, community stewardship of open space, lifelong learning about nature) are important attributes in realizing the full potential benefits of conservation techniques.

In new mixed-use developments, more public amenities, such as parks, trails, schools, and restaurants, make developments more attractive for homeowners. As these amenities are being built out in TNDs or conservation developments, there is a transition period during which development becomes more desirable for homeowners. After a minimum threshold of construction has been achieved, mixed-use developments outperform comparable conventional developments in price, sales velocity, and other critical factors.[100]

In the ten conservation developments analyzed for this book, open space is typically over 50% of gross acreage, compared to less than 20% in conventional design.[101] In a conservation development, storm-water management requires only a very small percentage of the open space. Typically, at least 30% open space is needed to allow homes direct visual, walking, and biking access to open space, which results in higher housing values. In conservation developments, open space commonly is protected from future development by means of a perpetual conservation easement.

A large body of research has found that property values increase with proximity to open space.[102] In green communities, open space is an important amenity for homeowners, who use trails daily or weekly, which may provide the additional benefit of increased physical activity and a decreased need to drive to obtain recreation. One study, in Washington County, Minnesota, found that homes near woods, fields, and lakes or streams are valued at $15,000 more than similar homes away from open space. Premiums for lots adjacent to open space ranged from 6% to 32% across various studies. A 1990 study in Massachusetts compared conventional developments to conservation developments that featured clustered housing and permanently preserved open space. On average, the homes in the conservation developments had 7,200-square-foot lots and sold for $137,000; in the nearby conventional developments, lots average 33,200 square feet and the homes sold for $102,000.[103] Cumulative appreciation for the conservation developments averaged 168% over eight years, with an average sales price of $367,000 at the end of the study period; in the conventional developments, cumulative appreciation was 147%, and the average ending sales price was $252,000.[104]

The Western Reserve Conservation & Development Council found that conservation developments were associated with higher lot premiums and more rapid absorption. At Laurel Springs, in Bainbridge, Ohio, lots abutting open space sold for 10% more than conventional lots, and lot absorption was at least 0.5 units more per month—leading to a 55% profit for the developer, versus 27% with conventional design.[105] At Thornbury, a development in Solon, Ohio, conservation lots sold for 10% more than conventional lots, and absorption rates were more than twice those for conventional lots (8 years to sell out versus 17 years).[106]

2.6. The Market Rediscovers Walkable Urbanism

CHRISTOPHER B. LEINBERGER

Throughout the 1990s, even as edgeless cities were pushing growth ever outward, the popular appeal of walkable urban life, whether in the cities or suburbs, was growing.[107] Although the changes were subtle at first, the 1990s witnessed a revival in many American downtowns—spurred in part, no doubt, by the dramatic drop in urban crime during that decade. New, walkable urban places were being developed in some suburban town centers; new development was occurring around transit stations; and new walkable development was being built from scratch on greenfields. These changes started slowly in the mid- and late 1990s, and took off in the 2000s. Much of this new development, which was generally located in suburbs, was sparked by new urbanism.

By the middle of the first decade of the 21st century, the country was moving in two diametrically opposed directions: metropolitan areas were expanding geometrically, as farms were converted into subdivisions named after what they had replaced—Whispering Woods, Bubbling Brook, Woodmont; at the same time, in a countertrend, downtowns were being revived and transit- and non-transit-served suburban town centers were taking off—bringing new development, revitalization, and excitement.

So what is it going to be over the next generation: continued low-density, drivable sub-urbanism; compact, walkable urbanism; or some combination of the two?

Demographic trends, consumer preferences, an emerging new version of the American Dream, and a recognition of the consequences of drivable sub-urbanism are all pushing the pendulum back toward walkable urbanism. Green community design will be an important part of the set of development practices that satisfy this large-scale shift in consumer demand.

PENT-UP DEMAND FOR WALKABLE URBANISM

The best evidence of the pent-up demand for walkable urbanism is the price per square foot consumers are actually paying for higher-density housing in walkable urban places, versus the prices they are paying for drivable sub-urban single-family housing in similar parts of the same metropolitan area. These data provide hard evidence that the market is willing to pay a significant premium for walkable urbanism.

In 2007, in Birmingham, Michigan, a walkable urban-suburban town in the economically depressed Detroit metropolitan area, the average per-square-foot price for a downtown condo (priced between $750,000 and $1,500,000) was $445.[108] A drivable sub-urban house in the same absolute price range—a few minutes away by car, but still in Birmingham—cost $318 per square foot. In the Detroit region—which, primarily because of its extreme dependence on car manufacturing, is one of the least walkable metro areas in the country—there is a 40% price premium for walkable urbanism.

In the Denver metropolitan area in 2007, luxury homes priced between $750,000 and $1,500,000 in Highland Ranch, a single-family master-planned community, sold for an average price of about $195 per square foot. However, if one wanted to enjoy the walkable urbanism that had exploded in downtown Denver over the previous decade, a comparably priced luxury home would have cost about $487 per foot, a 150% premium: two and a half times the price per square foot. High-income households seem to be willing to pay the same absolute dollar amount for a 4,000- to 7,000-square-foot suburban palace in a gated community that is near golf courses as for a downtown condominium that is about one-third the size but has city views and is within walking distance of the best selection of restaurants in the region—and maybe even work.

The New York City metropolitan area probably has the most extreme premiums for walkable urban housing in the country. In 2007, in wealthy Westchester County, north of New York, a price between $1 and $2 million for a drivable, single-family home translated into $365 per square foot. If you wanted the walkable pleasures of downtown White Plains, the major suburban city in the county, in 2007 you would have paid a 100% premium (twice the price per square foot, or $750) for a condominium. But if you wanted the excitement of Manhattan, it would have cost you, on average, $1,064 per square foot, a 200% premium: triple the per-square-foot price of a drivable single-family house in Westchester County.[109]

These examples suggest that housing prices in walkable urban places have between a 40% and a 200% premium over drivable single-family housing, controlling for price range and luxury orientation. Because the pent-up demand for walkable urbanism will probably not be fully met over the next 10 to 20 years, these price premiums will probably just increase.

An analysis by RCLCO compared the for-sale housing market in the Washington, D.C., metro area in 2006 (a year of housing weakness) and 2005 (the peak year of recent housing strength). The results showed relatively flat prices and a slight decline (12%) in sales pace for the District of Columbia and Arlington County, Virginia—both places with an abundance of walkable urban housing. In the far fringes of the D.C. metro area (Virginia's Loudon, Fauquier, and Prince William counties), where nearly all housing is drive-only sub-urban, sales prices were also relatively flat, but the sales pace had declined approximately 35%.[110] Drivable sub-urban housing on the fringe appears to have been most severely affected by the market downturn in the Washington region, but only time will tell for certain.

The disparity in prices signals that there is more demand for walkable urbanism than the real estate industry can produce. Because walkable urbanism is mainly illegal under current zoning, and is difficult to finance—and because the development industry does not yet fully understand how to create it—the supply has been insufficient. Unfortunately, however, there have been no definitive studies of the supply of walkable urban product. Atlanta and Phoenix, for example, may have no more than 10% of their housing supply in walkable urban neighborhoods. Older metropolitan areas, such as Boston and Chicago, may have 20% to 30%—or more—of their housing in walkable urban neighborhoods.

The preference for walkable urbanism is not confined to housing. In an analysis of the Washington, D.C., regional office market conducted by RCLCO, leases for walkable urban office space in late 2006 were 27% higher than those for drivable sub-urban space. Walkable urban space also had a much lower vacancy rate (7.7%) than drivable space (11.5%). The prime walkable urban office location is in downtown D.C., where office rents average $50 per square foot (some space is more than $60 per square foot), making this area the second-most-expensive office rental market in the country, after midtown Manhattan.[111] The prime drivable sub-urban office location is Tysons Corner, in Virginia, where office rents average $31 per square foot. In other words, office space in walkable downtown Washington, D.C., rents for a 61% premium ($19) over drivable sub-urban Tysons Corner.[112]

As a percentage of the total house price, land values in walkable urban places are much higher than in drivable sub-urban places. This imbalance in land values suggests how much money will be made in real estate as developers begin converting low-density suburban places into walkable urban places. When there is excess demand for one type of development, as is the case with walkable urbanism, the price of land spikes upward, causing windfall profits for some, and unaffordable housing for many. Over the next few decades, as the pent-up demand is gradually satisfied, the correction of the land-value imbalance will be *the* major market force affecting the real estate industry. Green, walkable community developments are exactly what is needed to satisfy the pent-up demand for walkable urbanism.

THE SHAPE OF FUTURE DEVELOPMENT

The built environment takes far longer to turn than the proverbial supertanker. In "The Longer View," Arthur C. Nelson notes that

> more than $30 trillion will be spent on development between the period 2000 and 2025. Nearly 50 million new homes will be built, including some 16 million that will be rebuilt or replaced

entirely with other land uses. Seventy-five billion square feet of nonresidential space will be built with 60 billion square feet replacing space that existed in 2000. New nonresidential development will equal all such development that existed in 2000.

Without planning, long-term thinking, and policy changes, the vast majority of this huge growth will be a continuation of drivable sub-urbanism. A 2007 survey of Urban Land Institute (ULI) members found that they believe there is a market for alternatives, but that municipal regulations remain the primary barrier to meeting this growing demand.[113] Long-term planning—including a shift to zoning laws that encourage green, walkable development—is essential if the United States, and its huge and crucial investment in the built environment, are to be properly positioned for the economic and environmental challenges of the 21st century.

Projections of the future of real estate prepared annually by ULI and PricewaterhouseCoopers concluded that

energy price [uncertainty] and road congestion accelerate the move back into metropolitan-area interiors as more people crave convenience. They want to live closer to work and shopping without the hassle of car dependence. Higher-density residential projects with retail components will gain favor in the next round of building. Apartment and townhouse living looks more attractive, especially to singles and empty nesters—high utility bills, car expenses and payments, and rising property taxes make suburban-edge McMansion lifestyles decidedly less economical.[114]

Confirming this perspective, the daily news site REBusiness Online recently reported that "developers are uniting this historically urban format with the increasingly popular 'live, work, play' motto of mixed-use development. The newly evolved transit-oriented development trend is taking root in suburban areas across the country."[115]

There are still obstacles to walkable urban development: skepticism about the depth of the unmet market demand; huge zoning impediments; Wall Street's reluctance to finance walkable urbanism, and the fact that drivable sub-urban product still gets significant subsidies. But even with all these obstacles, increasing numbers of walkable urban developments are being built. Just think about how much will be built when these hurdles have been removed, and the market can have what it wants.

2.7. Social Impacts of Green Communities

Neighborhood design that allows people to interact more easily with neighbors and to walk to multiple destinations increases the number of social contacts people have in a day, and leads to greater civic engagement and involvement in community organizations. These social impacts are generally most significant for teenagers and elders, who often lack access to vehicles or are unable to drive, and who consequently can be socially isolated in conventional, car-dominated communities. Residence times in TNDs and conservation developments also tend to be longer than in conventional developments, which helps make communities more vibrant: in conventional developments with larger lots and less shared open space, neighbors often don't know each other and are less likely to form lasting communities. Jane Jacobs, in her seminal work *The Death and Life of Great American Cities* observes that "the destructive effects of automobiles are much less a cause than a symptom of our incompetence at city building." Increased walking and social interaction in green designed communities reduces automobile dependence as an indirect consequence of placing resident and community needs at the center of city planning objectives.[116] Many recent green neighborhood developments have succeeded in creating communities that are diverse and engaged when compared with single-use residential developments, which often segregate both uses and people.[117]

The inclusion of a broader range of housing types in green communities, from large houses to small apartments, expands the range of prices and rents and increases the economic diversity of green neighborhoods. Similarly, by reducing household transportation costs, green community design effectively widens the definition of affordable housing. As noted earlier, the Center for Neighborhood Technology has developed an index that maps affordability by taking into account both housing and transportation costs. Moving from downtown to suburban Minneapolis, Minnesota, for instance, increases transportation costs from 10% to over 20% of income, even though income itself also increases.[118]

On the other hand, greening can bring about greater desirability and associated higher rents, making neighborhoods less affordable for renters with lower incomes.[119] The redevelopment of downtown areas as walkable, mixed-use developments—a green alternative to suburban sprawl—can lead to gentrification, and contribute to a shortage of affordable housing. Nevertheless, increased property values in previously depressed areas can be a boon to lower-income homeowners. In any green development, inclusionary zoning or other policies that help ensure adequate affordable housing are important tools for realizing social, financial, and sustainability goals.[120] (For a discussion of the costs and benefits of green affordable housing, see section 1.6, "Green Affordable Housing: Enterprise's Green Communities Initiative.")

Walkable communities that preserve or foster economic and social diversity provide many benefits—such as reduced transportation costs and improved health—that can be especially significant for low-income residents. The ability to meet basic needs, to socialize, and to get to work without a car is even more important for those without access to a car. A 2005 analysis of the results of the National Household Transportation Survey study found that facilitating pedestrian access to public transit may have the greatest health benefits for low-income populations.[121] Similarly, in conventional development, the adverse impacts of air pollution, lack of open space, and pedestrian-unfriendly environments disproportionately impact low-income populations.

RESEARCH AND CASE STUDIES

A 2001 study of Kentlands, a TND in Maryland, and Orchard Village, a neighboring conventional residential development, included a survey of 750 residents, in-depth interviews with 140 residents, and records of daily activities from 70 residents.[122] In the qualitative survey, Kentland residents showed more positive responses on four dimensions of community defined by the study: community attachment, walking, social interaction, and community identity.

Similarly, a study of Orenco Station, a transit-oriented community outside Portland, Oregon, revealed a high level of "social cohesion within the community." When compared with residents of two conventional communities, a higher proportion of Orenco residents reported that they had friendly neighbors and were more active in the community than in their previous residences. Residents also gave the physical design of the community higher satisfaction ratings than the residents of neighboring conventional communities.[123]

Two recent studies examined the social and physical impacts of the built environment on elders and youth in East Little Havana, a Hispanic neighborhood in Miami. One study examined the relationship between the physical health, mental health, and social interactions of 273 elders and the presence of so-called eyes on the street typical of traditional neighborhoods. The presence of front entrances and porches was found to be significantly and positively associated with elders' mental health and level of social interaction.[124]

Furthermore, a survey of the teacher-reported grades of students living in East Little Havana found that students living on a block with only residential buildings, with less access to commercial amenities, had 74% more reported problems with conduct (as indicated by being in the lowest 10% of conduct grades).[125] Structural disadvantages facing low-income communities (e.g., poor schools; difficulty accessing jobs and services; disproportionate burdens of pollution, disease, and crime) are often directly addressed by green strategies at the level of individual buildings or in green neighborhood design.

While more research is necessary to explore the dynamics of the relationship between walkable, mixed-use design and residents' mental, physical, and social health, a growing body of research indicates that a range of indicators of social cohesion and vitality are linked to green community design. Social impacts are difficult to value in financial terms, but it is likely that these impacts are already recognized in the form of increased property values: communities with greater social vitality become preferred places to live and work.

IMPLEMENTATION CHALLENGES

Design intent does not always achieve the expected benefits. This may be especially true in community-scale projects, where plans are often realized over the course of years, if not decades. For instance, Civano, a planned TND outside of Tucson, appears to have achieved many of its original green building goals while falling short of reaping the full benefits of a walkable, mixed-use design.

Civano was guided by strict standards for building design and neighborhood layout: walkable streets; front entrances; architectural standards requiring local building techniques and materials; water- and energy-efficiency goals; and a broad mix of uses, including single-family homes, live-work units, and commercial retail. A 2006 study found that Civano homes used 47% less energy and 57% less water than a typical Tucson household.[126] Subsequent stages of development, however, have been more conventionally suburban in form, in part because community groups and the original community planners have not succeeded in holding more recent developers to the original green standards. In addition, the town center has yet to be built, which limits residents' ability to reach amenities by walking.[127]

2.8. Cost Savings in Ecologically Designed Conservation Developments

CRAIG Q. TUTTLE, JILL C. ENZ, AND STEVEN I. APFELBAUM

For the last two decades Applied Ecological Services (AES) has worked with developers, municipalities, and corporations to create development that actively restores and protects natural ecosystems, habitat, and hydrology.[128] Conservation development extends other green development strategies, such as TND, by increasing the focus on restoring ecology and protecting natural open spaces. This section describes the results of a detailed cost comparison of conventional sprawl and conservation approaches in ten developments in the Midwest. The section also includes a discussion of the long-term financial impacts of conservation development.

Conservation development is an approach to suburban development that maintains far more open space than conventional development; it involves using smaller lot sizes, clustering buildings, creating more efficient development layouts, and using ecological restoration as a basis for design and development. An analysis of ten conservation developments conducted for this book demonstrated that even with additional costs for habitat restoration and landscaping, total on-site infrastructure costs were reduced 25% per home when compared with conventional sprawl. Even with 5% to 10% increases in the number of units per project, conservation developments still show project-wide infrastructure-cost reductions of 10%. Market data have shown that homes in conservation developments also appreciate faster than those in conventional suburban developments, and can command price premiums of 10% to 30%, even as lot size decreases and gross neighborhood density increases. Additional benefits include reduced storm-water flow and improved water quality and infiltration, increased wildlife habitat, increased access to trails and natural open space, and reduced long-term public infrastructure costs.

ECOLOGICAL CONSERVATION DEVELOPMENTS

Ecologically designed conservation developments start by identifying the land that is to be protected; the restoration of degraded lands, the creation of trails, and the use of alternative storm-water management then become major organizing elements in the design.[129] While conventional development may integrate some of these features, storm-water design and infrastructure are often integrated very late in the conventional design process.

The basic strategies used by AES, Inc., and other conservation developers include the following:

- Clustering housing
- Using smaller lots
- Creating distinct neighborhoods that are linked through green- and open-space systems
- Setting aside large portions of development footprint as natural areas
- Integrating a mix of uses
- Restoring open space and creating greenways and passive and active parks
- Creating extensive trail networks and ensuring that all homes have visual and/or pedestrian access to open-space systems
- Engaging the community in land management education and traditions
- Water management strategies, such as using land for storm-water management, using groundwater recharge to replenish the supply of potable water, reducing flood damage, mitigating flood-related downstream problems, and eliminating storm-water sewers and piped systems.

Ecologically designed conservation developments are sometimes assumed to be more expensive because much of the land (typically 50% to 60%) is set aside for open space (and is thus not subdivided into lots), and because of the additional cost of restoring that open space. There has also been concern over the marketability and price of smaller lots in conservation developments.

INITIAL COST COMPARISON

The figures presented here summarize the results of a comparison involving paired conventional and conservation plans for ten developments with a total of over 1,500 homes. The findings are based on detailed cost projections for each project, jointly prepared with the developers and with corporate and public clients of AES.

We used the following criteria to select the sample projects included in this analysis:

- The authors had experience with, and were involved in, the project.
- The project was a residential subdivision in the upper Midwest.
- The project had both a baseline conventional and an ecologically designed conservation concept plan.
- Sewer-service and water-supply requirements were the same for both concept plans.
- Both project designs met local ordinances.
- The project was a greenfield development.

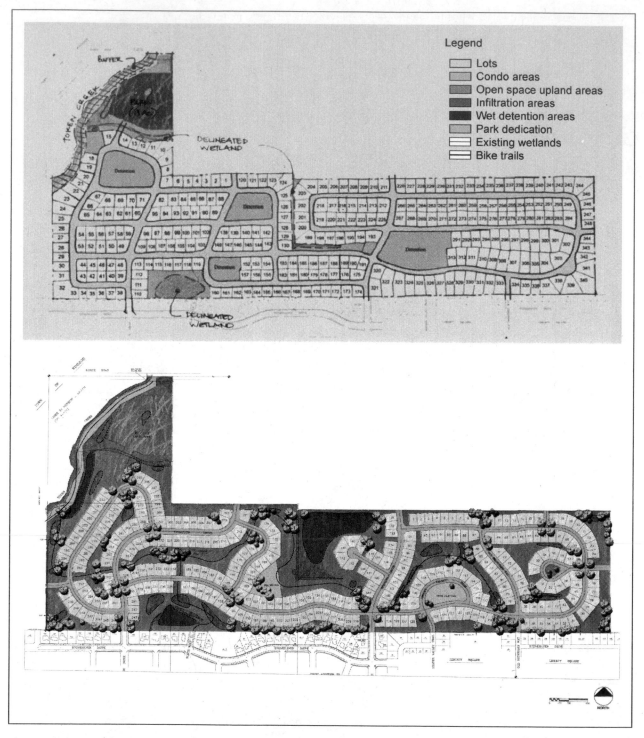

FIGURE 2.11 Conventional Development (top) and Conservation Development (bottom) Design Layout for Token Creek Conservancy Estates

Source: Applied Ecological Services, Inc.

- The project design was applicable to other geographic, regulatory, and market settings in the United States.

Under the conservation concept plan, most projects had over 50% of their total acreage in the form of open space. Conservation plans were designed in response to the specific hydrologic and ecological characteristics of each site, using alternative storm-water management principles and taking advantage of opportunities to restore streams, wetlands, native species, and habitat, and improve the aesthetic quality of open spaces.

SINGLE-DEVELOPMENT COMPARISON

Token Creek Conservancy Estates is a 203-acre development located on the northern edge of Sun Prairie, Wisconsin. There is a wetland at the south edge of the site, and Token Creek, designated as a Class III trout stream by the U.S. Department of Natural Resources, serves as the western boundary of the site. This creek is the largest tributary to Lake Mendota, which borders Madison, Wisconsin.

TABLE 2.5 TOKEN CREEK COST COMPARISON

Cost category	Conventional development costs ($)	Conservation development costs ($)	Absolute difference ($)	Difference in percentage terms
Grading	1,425,418	947,142	478,276	−34%
Roadways	2,313,896	1,512,412	801,484	−35%
Storm sewers	1,145,639	519,544	626,095	−55%
Sanitary sewers	1,502,840	1,105,282	397,558	−26%
Water main	1,657,739	1,233,850	423,889	−26%
Erosion control	35,684	35,684	0	0%
Off-site sanitary	26,250	26,250	0	0%
Landscaping/ restoration	284,200	665,192	−380,992	134%
Amenities	999,222	732,240	266,982	−27%
Contingencies (engineering, legal)	2,347,722	2,347,722	0	0%
Project total	11,738,610	9,125,318	2,613,291	−22%
Per-unit total	38,237	26,839	11,397	−30%

TABLE 2.6 CONVENTIONAL VERSUS CONSERVATION DEVELOPMENT: A COMPARISON OF MULTIPLE DEVELOPMENTS

Project name	Project location	Acres	Number of units: conventional	Number of units: conservation
Laurel Springs	Jackson, Wisconsin	42.5	112	126
Greenview Meadows	Sullivan, Wisconsin	30	22	22
Shore Ridge	Walworth County, Wisconsin	120.78	21	35
Friedrich	Ames, Iowa	180.8	120	184
Copperleaf	Kansas City, Missouri	198.22	444	313
Inspiration	Bayport, Minnesota	242.4	302	358
Auburn Hills	Germantown, Wisconsin	81.75	133	126
Rountree Branch	Platteville, Wisconsin	41	46	70
Rolling Hills of Hammond	Town of Hammond, Wisconsin	140	62	77
Token Creek Conservancy Estates	Sun Prairie, Wisconsin	203	307	340

Note: Data on completed projects suggest that the projections are borne out in actual construction costs experienced by AES on these and other projects. Because the projections from uncompleted projects included in this multiple development comparison can therefore be assumed to be relatively accurate, they have been included along with the completed projects.

Both the conventional and conservation concept plans described here assume city sewer and water. The conventional design accommodates 307 lots ranging from 15,000 to 25,000 square feet, while the conservation design accommodates 314 quarter-acre lots (10,890 square feet) plus 30 townhouse units. Figure 2.11 shows the conventional and conservation design plans.

Storm-water management in the conventional plan consists of a typical curb and gutter network, with curb inlets and pipes running into a few detention basins on site. The conservation concept uses a patented method known as a storm-water treatment train (STT), which consists of open prairie, rain gardens, bioswales, and wetland biofiltration cells—natural systems that remove sediment, phosphates, nitrogen, and other pollutants, and contribute to comparatively higher water quality. The STT meets or exceeds requirements for storm-water management—but, more importantly, because it promotes infiltration and groundwater recharge, the STT maintains greater water flow to Token Creek and reduces the adverse thermal impact of storm-water runoff on cold-water trout habitat (infiltrated water is naturally cooled before entering the creek).

TABLE 2.6 *(CONTINUED)* ─────────────

Build status	Cumulative cost savings ($)		Percent change in costs	
	Per project	**Per unit**	**Per project**	**Per unit**
Under construction	386,503	6,238	−12%	−22%
Under construction	52,789	2,399	−5%	−5%
Under construction	−37,422	51,139	1%	−39%
Not constructed	−732,508	12,091	13%	−34%
Under construction	4,783,964	2,082	−34%	−7%
Under construction	2,028,622	10,876	−20%	−33%
Under construction	625,613	2,887	−13%	−16%
Not constructed	29,359	11,890	−2%	−36%
Under construction	264,373	17,934	−6%	−24%
Under construction	2,613,291	11,397	−22%	−30%
Average percent change in cost with conservation development	**1,001,459**	**12,839**	**−10%**	**−25%**

The ecological conservation design results in a savings of 22%, or $2,613,291, development wide, and 30%, or $11,397, per unit (see table 2.5). The primary sources of savings in this particular project are the storm-water systems (55% reduction; $626,095 saved); roads (35% reduction; $801,484 saved); and grading (34% reduction; $478,276 saved). More money was spent on landscaping in the ecological development, however: an additional $381,000 for restoration. The savings were realized even when this additional expenditure was taken into account.

Restoration in conservation developments includes planting native species, developing trails, and using bioswales and restored habitat to infiltrate water, minimize erosion, settle solids, absorb nutrients, and reduce the velocity and quantity of runoff entering rivers and downstream wetlands.[130] Such improvements provide ecological benefits, such as improved wildlife habitat and reduced storm-water pollution, as well as amenities for residents, who can enjoy increased access and connection to natural open space. The use of native and restored vegetation also decreases the need for irrigation and pesticides when compared with landscaping in conventional developments.

TABLE 2.7 CONVENTIONAL VERSUS CONSERVATION DEVELOPMENT:
AVERAGES FOR TEN PROJECTS

	Conventional	Conservation
Number of residential units	157	165
Open space	14%	59%
Roadway length (feet)	122,021	96,691

MULTIPLE-DEVELOPMENT COMPARISON

A comparison of ten AES projects, with a total of over 1,500 residential units, shows that ecological design yields average cost savings of 10% development wide and 25% per home (see table 2.6). The difference in the savings percentage arises from the fact that in seven of the ten projects studied, the conservation plan calls for more homes than the conventional plan.[131] Despite an overall increase in the number of units, the conservation plans average 59% open space, versus 14% open space in the conventional plans. Eight out of ten projects show overall cost savings for conservation development; more importantly, all ten show savings on a per-home basis.

The primary areas of savings for conservation development in these ten examples are reduced storm-water systems (39% average cost reduction); reduced roadways (18% average cost reduction); and reduced grading (39% average cost reduction). On all conservation projects, more money was spent in the landscaping and restoration category (147% average cost difference). Based on the 1,500+ units in these ten developments, average *net* cost savings for conservation development are over $12,000 per unit.

Site preparation and grading, in particular, can be the hardest projected costs to pin down. With this in mind, the cost comparison was run a second time, with an alternative-grading scenario in which greater costs were attributed to topsoil stripping and lot grading. Under this scenario, the average cost savings with conservation development increased to 16% project wide and 29% per unit.

Overall, a comparison of the ten projects also showed that conservation developments were characterized by significantly more units, smaller lot sizes, more open space, and shorter roadway lengths (see table 2.7). These design changes have significant long-term benefits: lower public infrastructure costs (in large part from reductions in storm-water flow), higher property value, more wildlife habitat, and more public land for recreation.

2.9. International Green Building

Buildings account for about 40% of primary energy consumption in most countries. Buildings' share of global CO_2 emission varies significantly between countries, and is estimated variously at 20% to over 50%, depending on the method of emissions accounting.[132] The International Energy Agency estimates that energy demand from buildings will drive half of new energy supply investments up until 2030.[133]

A number of countries have used government directives and building codes to encourage or require energy efficiency or health standards for buildings. These range from appliance-efficiency and labeling rules, to mandated utility-efficiency programs, to required energy-performance documentation for commercial buildings.[134] In past decades, the United States has generally pursued a voluntary approach toward incentives for energy efficiency, while Europe has introduced mandatory energy- and water-efficiency requirements for new buildings.

As a result of differing policies, architectural practices, and climates, and variations in building space per person, building efficiency varies dramatically between countries. As countries develop, the amount of building space per person generally increases, along with the portion of building energy use made up by electricity. The United States has over 850 square feet of building space per person; the European Union (EU) has 550 to 650 square feet, and China has less than 350 square feet. As electricity or natural gas replaces fuels burned on site (e.g., wood, coal, biomass), indoor air quality and energy efficiency generally improve. However, the use of electricity typically increases primary energy consumption and CO_2 emissions, unless that electricity comes from low- or zero-emissions sources. Future global energy use and CO_2 emissions from buildings depend heavily on whether countries like China, India, and Brazil follow the wasteful and high-consumption patterns of the United States or the more efficient patterns of Europe and Japan. From the perspective of energy use and climate change, these alternatives are substantially different; for instance, if China follows European consumption patterns, its energy use by 2050 may be less than three times what it is today; if China follows U.S. consumption patterns, its energy use by 2050 may be more than five times what it is today.[135]

Rob Watson, *Chief Executive Officer, EcoTech International*

Every year, China erects over 2 billion square meters (21.5 billion square feet) of floor area.[1] The energy and resources required to build and operate that much space take a tremendous toll on the environment, creating impacts on the landscape, water resources, and energy use; causing deforestation; and affecting human health and productivity.

Each year, China's building boom covers nearly a million hectares (2.4 million acres) with concrete and asphalt, of which roughly 20% was formerly productive farmland—a resource that China can ill afford to squander. When green land is turned into hardscape, groundwater can't regenerate, which slows the replenishment of the aquifers that supply 70% of China's drinking water and 40% of its agricultural water.

Eighty percent of China's potable water is used in buildings, requiring the development of a vast municipal infrastructure to capture, store, purify, distribute, and treat water and sewage. Buildings command almost 30% of China's energy use (45%, including embodied energy). In large urban areas, more than half the peak demand comes from air conditioning and lighting, which is driving the construction of one new power plant every week or two.

In the 2001 Yangtze floods, China learned hard lessons about uncontrolled timber harvesting. Since then, a massive reforestation program and controls on logging in western Sichuan Province have successfully begun to reverse this damage. Unfortunately, continued demand for wood, largely driven by demand for furniture and finishes, has forced unsustainable logging practices off shore, principally to Southeast Asia and Canada.

As China becomes more urban and the economy becomes increasingly service oriented, people are spending more and more time indoors—where pollution from cigarette smoke, cooking, and the chemicals used in furnishings, carpets, and finishes results in a noxious soup that can be especially harmful to infants and elders. Poor indoor air quality, substandard lighting, noisy acoustical environments, and poor temperature control can reduce economic productivity by up to 20%, and contribute to approximately one-third of health-related absenteeism in the knowledge-related and service sectors.

Fortunately, green building in China has grown rapidly in recent years, and has the potential to mitigate each of the environmental impacts just described. Through its Office of Energy Efficiency in Buildings, China's Ministry of Construction (MOC) offers a green building label based on the 2006 Evaluation Standard for Green Building, which combines elements of CASBEE (the Japanese system), and the U.S. Green Building Council's Leadership in Energy and Environmental Design rating system.

Over the last few years, China has adopted and begun implementing several policies and programs promoting energy-efficient and green buildings. In addition to the Evaluation Standard for Green Building, the MOC has promulgated energy-efficiency standards for new construction that require at least a 50% improvement over 1980s levels of energy efficiency. In major cities—Beijing, Shanghai, and Guangzhou—new buildings must reduce energy consumption by 65% in relation to 1980s levels. These standards apply to both residential and commercial properties and will bring minimum performance requirements into line with U.S. standards, but are still well below European standards.

Implementation and enforcement of the standards remain a challenge in many areas. Though "heaven is high and the emperor is far away," rigorous enforcement is becoming more prevalent, and heavy fines have been levied for noncompliance. The MOC may also revoke a developer's license for repeated noncompliance. Nevertheless, design expertise and energy-saving materials, technology, and availability still lag behind China's rampaging development market.

The Ministry of Finance and the National Development and Reform Commission have approved several provinces to begin collecting funds to support market-based incentive programs,[2] including incentives and preferential policies for certain types of equipment. Beijing and Xinjiang, for example, provide incentives for the installation of ground source heat pumps in new projects. In some cases, the incentives are financial and linked to the size of the system; other places use accelerated permitting and development approval to encourage the use of new technologies.

---

1. EcoTech International implements clean-tech solutions to integrate sustainability over the life of buildings. Watson was the founding chairman of LEED, and has worked for over a decade to promote green buildings in China.
2. To provide Chinese-language information on green building in China, the MOC supports the China Green Building Network Web site: www.cngbn.com. The TopEnergy bulletin board (http://bbs.topenergy.org/) is an excellent professional networking resource.

Green building has been spreading across both developed and developing countries in recent years. Green building standards have been developed independently in a number of countries; the list includes the British Research Establishment Environmental Assessment Method (BREEAM);[136] Green Star, in Australia;[137] the Hong Kong Building Assessment Method (HK-BEAM);[138] and the Comprehensive Assessment System for Building Environmental Efficiency (CASBEE), in Japan.[139] LEED, which is the most widely used green standard in the world, has been applied to projects in nearly 100 countries. Despite differences between the standards and the countries in which they were developed, all tend to recognize a very similar range of building attributes, including site environmental impact, energy and water efficiency, indoor environmental health, and the use of sustainable materials.

The World Green Building Council (World GBC), founded in 2002, supports the creation of green building councils in countries around the world. As of February 2009, the World GBC had 13 country members with active national green-building rating systems, and ten "emerging" member councils that are developing national systems. Member countries alone represent over 50% of the global construction market.[140] The World GBC does not promote a single international standard, but works with members to facilitate the creation of national standards that respond to local constraints and conditions.

LEED registrations for non-U.S. projects have accelerated noticeably since 2006, more than tripling each year. According to USGBC records, 85 non-U.S. projects were registered in 2006, 353 in 2007, and over 1,139 in 2008.[141] In 2008, across all LEED standards, non-U.S. buildings represented 14% of new registrations and over 30% of all newly registered space. When non-U.S. buildings seek certification from the USGBC, they are required to meet all requirements, including those based on U.S. benchmarks such as ASHRAE standards, unless the applicant can demonstrate that local standards are of equal or greater stringency.

Perhaps the most aggressive international efficiency standard, the German Passivhaus designation, requires energy use of no more than 13,000 Btu/sf (for central European climates). By contrast, average U.S. residential energy use in 2005 was roughly 44,000 Btu/sf.[142] Even in cold climates, the best Passivhaus buildings can almost eliminate the need for heating systems, relying instead on super-tight construction and insulation to retain the heat given off by occupants and indoor appliances.[143] Over 7,000 homes in Europe have been built following Passivhaus principles.[144]

Twelve non-U.S. green buildings are included in the data set for this book (see appendix C). These projects are located in Australia, Canada, Dubai, Finland, Germany, and Mexico, and have used a number of different green building standards, including LEED and Australia's Green-Star program. The buildings show a similar range of cost

TABLE 2.8 INTERNATIONAL GREEN BUILDINGS

Project name	Country	Building Type	LEED level or equivalent	Year constructed
30 The Bond	Australia	Office	5-star green star rated	—
40 Albert Road	Australia	Office	—	2004–05
Bordo International	Australia	Office	5-star green star rated	2004
CII–Sohrabji Godrej Green Business Center	India	Office	Platinum	2003
Center for Sustainable Building – ZUB	Germany	Office	—	2002
C. K. Choi Building	BC, Canada	Higher education	Gold equivalent	1996
Conservation House	New Zealand	Office	5-star green star rated	2007
Eco-vikki	Finland	Residential community; variety of housing types	—	2004
Fire and Emergency Services Training Institute, Greater Toronto Airport Authority	Canada	Institutional facility	Silver	2007
Jeunes Sans Frontières Secondary School	Ontario	High school	Silver	2007
PAPSA Monterrey	Mexico	Office	Gold	2007
Paul Wunderlich-haus	Germany	Office	—	2007
PCL Centennial Learning Centre	Alberta	Office	Gold	2006
University of British Columbia, Life Sciences Centre	BC	Higher education	Gold	2008
The Wafi City District Cooling Chilled Water Plant – DCCP ONE	Dubai	District cooling plant	Gold	2006

Note: Cells marked with a dash (—) indicate that the information was not available.

premiums as U.S. green buildings: from 0% to 12% more than conventional construction. Reported energy-use reductions range from 20% to over 70% (see table 2.8). Green buildings and green design are growing very rapidly globally with some counties already ahead of the United States. The EU has set a target of 2019 for all new residential construction to be zero net energy, a year ahead of California and Massachusetts targets. National green design programs are broadly similar but each is commonly tailored to

TABLE 2.8 *(CONTINUED)*

Green premium (cost difference between actual building and the same building constructed using conventional building practices, including hard and soft costs) as a % of total building cost	% Reduction in energy use compared to conventional building	Energy baseline used for calculating % reduction	% Reduction in water use compared to conventional building
—	—	—	—
—	—	—	—
12%	68%	Typical Australian office	0%
—	55%	—	30%
—	77%	Typical German office building	—
0.0%	57%	ASHRAE 90.1	—
—	40%	—	60%
<5%	20%	Local buildings of same age	22%
1.5%	31%	MNECB	20%
4.5%	47%	MNECB 1997	31%
12%	29%	ASHRAE 90.1	20%
0.00%	60%	Typical European Building	0%
—	38%	Model National Energy Code for Buildings (MNECB)	43%
0.4%	30%	ASHRAE 90.1 1999	9.65%
4.5%	30%	ASHRAE 90.1 1999	55%

local climate conditions and increasingly encourages indigenous efficient and green design approaches and materials.

Globally, green design is becoming a core branding strategy for a growing number of developments and resorts such as the zero net energy Playa Viva development in Mexico that includes community ecological investments such as mangrove restoration and expansion of an on-site turtle hatchery. Global real estate developers such as Redevco are

Paul King, Chief Executive Officer, U.K. Green Building Council

The global financial crisis has presented an unexpected opportunity—an opportunity to make sure that the economic recovery is a low-carbon one, which puts the climate at center stage. Not just in the U.K., but around the world, in both developed and developing nations.

But how are we going to do this? The challenge is twofold. Firstly, stop the rot. New buildings need to emit "zero carbon"—ASAP. Secondly, and the greater challenge in developed countries, is to refurbish our existing homes and buildings so that they go on a radical and rapid carbon diet.

Over 95% of our U.K. members—private sector organizations not traditionally associated with a love of regulation—in a recent survey said during the credit crunch that it was very important that government should stick to its ambitious green targets, thus providing certainty on the direction of future policy. The U.K.'s carbon budgets, including recommendations made by the Climate Change Committee in December 2008, are an important way of achieving this.

Capital expenditure by governments can make or break a "Green New Deal." Spending should go to public transport, skills, energy efficiency, and public sector building programs with high sustainability standards. But if it also goes to new road-building programs or increased airport capacity, as will probably happen in the U.K., economic activity will perpetuate conventional, unsustainable models of building.

In practice, a new green deal means more energy-efficient building fabric and appliances, a mixture of small-scale and local renewable solutions, and a greener large-scale energy supply throughout the grid. At every stage, an injection of capital investment now in these areas will boost growth, jobs, and savings.

Clearly there is reluctance to truly let go of the old way of thinking, but the surge of interest in green building in the U.K. and elsewhere is evidence that this is beginning to change. A new international climate change deal that includes the United States, China, and India—along with national rescue packages that incentivize a cleaner, greener way of doing business—would be a promising start.

greening their construction even in markets like Turkey where demand for green is only beginning to emerge.

Greening of construction in China and India will have a far larger impact on global energy use and climate change than what happens in the North American or European building markets. And some developing countries, including China and India, are experiencing extremely rapid growth in green design so that within several years they could have more green space than either the US or Europe, potentially making them global leaders in developing, deploying and exporting green technologies.

2.10. Financial Impact of Green Communities

The financial benefits of green community design are significant, and are likely to far outweigh the additional development costs. Further research is needed to estimate the values of specific anticipated outcomes for various approaches in different settings, but the general financial impacts are clear, and typically include the following:

- For residents, homeowners, owners, and renters of commercial property, choosing to locate in green communities may entail higher initial purchase prices or rents. Long-term benefits will commonly include reduced transportation costs; reduced health costs; higher resale prices for property; improved pedestrian safety; greater social vitality; and, commonly, greater contact with nature.
- For developers, green community development may entail increased risk because of the complexity of design and higher construction costs. Benefits include potential reductions in on-site infrastructure costs, less opposition from community groups (that is, from groups that oppose conventional development), and, over time, increased property values for both land and buildings.
- For municipalities, an increase in costs for public transit and potential increases in costs for maintaining public areas will be accompanied by substantially reduced per-unit infrastructure costs; reductions in traffic congestion, carbon emissions, storm-water runoff, and pollution; a greater ability to accommodate future growth; and an increase in long-term economic activity and tax base.

Many of the benefits of green community design are difficult to quantify, in part because of the large variety of green community types. Table 2.9 summarizes four benefits of green community design for which we have developed financial estimates. Although estimating the magnitude of many of the costs and benefits of green community design is beyond the scope of this book, these impacts are significant and should be considered by developers, planners, and governments involved in community-scale developments. Additional benefits, including habitat protection; reductions in long-term infrastructure costs, storm-water runoff, and emissions; and increases in property

TABLE 2.9 BENEFITS OF GREEN COMMUNITY DESIGN

Benefit	Annual benefit estimate (per household)	Present value[1]		Basis of estimate
		Per household	Per square foot	
Lower up-front infrastructure costs	First-cost savings	$12,000	$5	Analysis of ten conservation developments
Lower transportation costs	$1,700	$21,000	$9	Comparison of transportation patterns in walkable and car-dominated neighborhoods in Atlanta. Estimates of average transportation cost from reductions in the amount of gas used, assuming that 25% of walkable households eliminate the ownership of one car.
Lower health costs	$300	$3,600	$2	Comparison of physical activity in walkable and car-dominated neighborhoods in Atlanta. Estimates of health costs based on level of physical activity.
Reductions in car accidents and injuries	$200–$300	$3,000–$4,000	$1–2	National data on costs of accidents and injuries; research on accident/injury rates in narrow, walkable streets. Estimate assumes 50% reduction in crashes in green communitites due to narrow, walkable streets.

1 Calculated over 20 years, with a 7% discount rate.

values and community vitality appear material and are important areas for further research. If energy prices increase, the pent-up demand for green development is likely to increase as well—meaning that in the future, green community developments would likely enjoy additional increases in value.

PART III

COMMUNITIES OF FAITH BUILDING GREEN

More and more communities of faith—including Protestant, Catholic, Jewish, Muslim, Buddhist, and Quaker groups— are embracing green design and green building. While beliefs, traditions, and practices vary in many respects, care for the earth is a value that transcends religious distinctions and emerges as a common motivation for incorporating environmentally friendly designs into construction projects. A study of 17 faith-based organizations that have recently invested in green buildings reveals a common sense that building green is a way of committing an entire community to the moral imperative to care for the earth and help all people share in the benefits of a healthy, sustainable environment.

Belief in a higher being, respect for creation, and a mandate to care for one's neighbor are at the core of many faiths. Across the entire religious spectrum, more and more people of faith are articulating a spiritual response to environmental degradation. Many religious traditions call upon members to be good stewards of the earth and its resources. Caring for one's neighbor means actively protecting the environment on which that neighbor's life and health depend. Religious communities are conscious of serving as the repository of universal and long-held values: duties to the oppressed everywhere, and obligations to future generations and to the whole of creation. In other words, religious communities are committed to advancing "the planetary common good."[1] As a fundamental religious and moral priority, the call to care for our planetary environment can serve as a unifying perspective within and between religious traditions.

3.1. Faith Groups in the Green Vanguard

Today, the ecological crisis has assumed such proportions as to be the responsibility of everyone. . . . I wish to repeat that the ecological crisis is a moral issue. . . . As a result [Christians] are conscious of a vast field of ecumenical and interreligious cooperation opening up before them.

—POPE JOHN PAUL II, *1990 World Day of Peace Statement*

Over the past two decades, the moral responsibility to care for the earth has come to pervade authoritative documents and public declarations by major faith groups. A scan of faith-based educational materials, faith-based journals, national faith-based and environmental Web documents, and the popular press reveals that environmental concerns are affirmed in mission statements, educational materials, and national campaigns sponsored by widely respected religious sources. In 1990, the same year that the late Pope John Paul II made the statement quoted in the epigraph, the World Council of Churches declared, "Today, all life in the world, both of present and future generations, is endangered. . . . The magnitude of the devastation may well be irreversible and forces us to urgent action."[2] Speaking for Greek Orthodoxy internationally, Ecumenical Patriarch Bartholomew I declared, "For humans to cause species to become extinct. . . . to degrade the integrity of Earth by causing changes in its climate, by stripping the Earth of its natural forests . . . to contaminate the Earth's waters, land, air and life with poisonous substances—these are sins."[3]

Fifteen years ago, in an extraordinary demonstration of the breadth of accord in the American Jewish community, the most senior officers of the Union of Orthodox Jewish Congregations of America, the Central Conference of American Rabbis, the Rabbinical Council of America, the Union of American Hebrew Congregations, the United Synagogue of America, and the Conference of Presidents of Major American Jewish Organizations established the Coalition on Environment and Jewish Life. In its founding statement, the coalition declared as follows:

> We, American Jews of every denomination, from diverse organizations and differing political perspectives, are united in deep conviction that the quality of human life and the Earth we inhabit are in danger, afflicted by rapidly increasing ecological threats: . . . global warming, massive deforestation, the extinction of species, poisonous deposits of toxic chemicals and nuclear wastes, and exponential population growth. As heirs to a tradition of stewardship that goes back to Genesis . . . we cannot accept the escalating destruction of our environment and its effect on human health and livelihood.[4]

In a declaration first distributed by the highly regarded relief and development agency World Vision, nearly 500 prominent evangelical scholars and agency executives agreed:

> We and our children face a growing crisis in the health of the creation in which we are embedded, and through which, by God's grace, we are sustained. These degradations can be summed up as: 1) land degradation; 2) deforestation; 3) species extinction; 4) water degradation; 5) global toxification; 6) alteration of the atmosphere and 7) human and cultural degradation.[5]

In March 2008, over 50 pastors and leaders from the Southern Baptist Convention released a call to action on the environment, declaring that "it is time for individuals, churches, communities and governments to act."[6]

Drawing on rigorous, peer-reviewed scientific consensus, faith groups have reached certain conclusions about the urgency of environmental conditions that raise fundamental religious and moral concerns and call for appropriate responses. Catholic bishops have consulted their own Committee on Science and Human Values, and have reviewed scientifically authoritative reports from sources including the Intergovernmental Panel on Climate Change. Jewish leaders have drawn on reports from the National Academy of Sciences. Eastern Orthodox officials have deliberated with representatives of the American Academy for the Advancement of Science.

On Earth Day 2009 (April 22), the U.S Conference of Catholic Bishops (USCCB) and the Catholic Campaign on Climate Change announced a new wave of environmental initiatives related to climate change, including a proposal that each of the nation's 19,000 Catholic parishes undertake action and advocacy; the first step in that direction will be an assessment of the local carbon footprint of each parish. Since 1993, the conference's Renewing the Earth environmental-justice program has provided resource materials, developmental support, conferences, retreats, and small seed grants for dioceses, parishes, schools, universities, and regional and local groups affiliated with environmental education, advocacy, and activism.[7]

Response to moral concerns about the future of the environment is strengthened by ecumenical collaboration. For example, in 1993, the USCCB, the National Conference of Churches of Christ, the Evangelical Environmental Network, and the Coalition on the Environment and Jewish Life formed the National Religious Partnership for the Environment (NRPE). In addition to undertaking scholarly and public policy initiatives, the NRPE provides resources to over 135,000 congregations around the country on local environmental actions and ways to incorporate the theme of environmental responsibility into religious worship and study.

Religious communities have also linked environmental concerns to issues of social and economic equity. In 1991, the USCCB wrote that "the ecological problem is intimately connected to justice for the poor."[8] Largely as a result of direct ministry to the planet's most vulnerable peoples, organizations such as Catholic Relief Services, the Church World Service, and the Association of Evangelical Relief and Development Organizations are increasingly recognizing the importance of further integrating efforts for environmental sustainability into their long-standing efforts to improve health, nutrition, and life expectancy. The NRPE has sponsored activities relating to the impact of environmental degradation on the poor and on racial minorities in brownfield areas, and along the U.S. border with Mexico; has supported various initiatives promoting children's environmental health; and has been a leading advocate for the protection of coastal and island nations affected by climate change. With the late Pope John Paul II, the members of the NRPE understand that "the right to a safe environment" is a fundamental right for all peoples.[9]

To help reduce the incidence of asthma, cancer, birth defects, learning disabilities, and other health problems caused at least in part by environmental influences, the National Council of Catholic Women, the Catholic Health Association, the National Catholic Education Association, and the U.S. Catholic Conference's Secretariat of Pro-Life Activities have established the Catholic Children's Health and Environment Campaign. To mitigate climate change and air pollution, the National Council of Churches of Christ mailed educational packets to almost 100,000 congregations explaining how best to conserve energy.

According to the NRPE, faith communities have organized to protect threatened local areas, often working to relieve the excess burdens of pollution and related health problems borne by poor and minority communities.[10] For instance, churches in the Catholic Archdiocese of Santa Fe are addressing the interaction of poverty, racism, and environmental degradation along the Rio Grande corridor.

Green building, renewable-energy, and energy-efficiency projects provide opportunities for religious institutions and communities of faith to take tangible steps toward reducing their own impact on the environment, leading by example to encourage a national shift toward a more sustainable built environment. Four recent monastic initiatives follow a millennium-long tradition of dwelling in harmony with the land, and demonstrate a conviction that environmental responsibility includes enhancement of human habitat, including its aesthetic dimensions.[11]

- The Sisters of the Presentation, in Los Gatos, California, undertook a green renovation of their retreat in the Santa Cruz Mountains.[12]
- Benedictine monks at the Abbey School, in Portsmouth, Rhode Island, have undertaken a variety of green projects, including the installation of a 660-kW wind turbine.[13]
- Sisters at the Sacred Heart Monastery, in Richardton, North Dakota, installed two wind turbines on their site, which allow them to save over 40% on their energy bills.[14]
- The Benedictine Sisters of Perpetual Adoration, of Clyde Monastery in rural Missouri, are building a 289-foot wind turbine on their property as part of a local energy cooperative.[15]

3.2. Methodology and Findings

Wherever they live, work, and worship, people of faith have established diverse initiatives to respond to the moral imperative for stewardship of the earth. For this section of the book, we assembled a cross-section of 17 faith-based communities and institutions that have recently invested in green building initiatives. Through detailed surveys and interviews, we examined the motivations and decision-making processes associated with building green in these communities. Several initial questions informed the approach to this research:

- Are green building initiatives in faith communities motivated by the same financial and environmental cost-benefit considerations described in the rest of the book, or by other less quantifiable, but no less important, factors?
- What additional challenges or benefits are experienced by communities of faith that decide to build green?

An interview instrument was developed to document the experience of faith groups in the organization, implementation, and assessment phases of green building projects. (The instrument is included in appendix K.) Internet searches, including a review of the USGBC's lists of LEED-registered and -certified projects, were used to develop a list of green building projects sponsored by faith-based organizations; the initial search generated a list of 40 green building projects. Each faith-based organization affiliated with these projects was contacted, and a lead respondent was identified. Data were collected through multiple interviews, and through e-mail correspondence with congregational and community leaders, facilities directors, committee chairs, architects, engineers, public-relations officers, and faculty. Information from archival and other data sources (e.g., mission statements, case studies, articles describing the projects) was collected and analyzed. When possible, supplemental data were collected on the costs and financial impacts, including energy and water savings, experienced as a direct result of new green construction; these financial and building performance data are included in appendix C.

Detailed responses were obtained from 17 faith-based organizations. The green building projects that respondents had undertaken included places of worship, offices for national or community-based organizations, classrooms and academic buildings on the campuses of faith-based universities, a convent, an elementary school, and a community center. All projects were completed or had projected completion dates between 2002 and 2008, and all were certified or anticipating certification under LEED, or were built with equivalent green goals.

Table 3.1 (pages 156 to 159) lists the 17 projects and includes brief descriptions of the motivations and impact of the projects for each community. Nine of the buildings were able to supply data on cost and energy and water savings. Green premiums ranged from 0.6% to 9.6%, with a median of 3.4% (higher than the less than 2% median for the larger data set). Reported energy savings ranged from 22% to 60%, and reported water savings ranged from 2% to 93% when compared with conventional buildings.

As can be seen in table 3.1, there are common threads in the motivations for undertaking green building projects in faith communities, the impacts green projects have had on faith communities, and the dynamics of the community decision-making and implementation process. For more information, please see www.islandpress.org/Kats.

ORGANIZATIONS

Project	Religious affiliation	Green standard; year achieved or anticipated	Location	Data source
Catholic Relief Services (CRS) Headquarters	Catholic	LEED certified, 2007	Baltimore	Dave Piraino, executive vice president of human resources
Capitol Hill Building, Friends Committee on National Legislation (FCNL)	Quaker	LEED Silver, 2005	Washington, D.C.	Maureen Brookes, communications program assistant
Community Center, Muslim Khatri Association (MKA)	Muslim	Environmental Review Certification: Gold Star Award, 2002	Leicester, England	Yahya Thadha, center manager
Headquarters, American Jewish Committee	Jewish	LEED Silver (anticipated), 2008	New York, NY	Ben Tressler, green building manager
Convent and high school, Felician Sisters of Pennsylvania	Catholic	LEED Gold, 2004	Coraopolis, Pennsylvania	Sr. Mary Christopher Moore, provincial minister

SCHOOLS AND ACADEMIC BUILDINGS

Project	Religious affiliation	Green standard; year achieved or anticipated	Location	Data source
Harm A. Weber Academic Center, Judson University	Evangelical	LEED Silver, 2006	Elgin, Illinois	Tonya Lucchetti-Hudson, director of communications
Morken Center for Learning and Technology, Pacific Lutheran University (PLU)	Lutheran	LEED Gold, 2006	Takoma, Washington	Rose McKenney, professor of environmental studies and geosciences; chair of Campus Sustainability Committee
Campus Center, University of Scranton	Catholic	LEED certified (anticipated)	Scranton, Pennsylvania	Springs Steele, associate provost for academic affairs and chair of the Sustainability Task Force

TABLE 3.1 *(CONTINUED)*

Motivation	Impact
(1) To achieve greater solidarity with the world's poor through more equitable and responsible use of the earth's resources, which reflects principles that are inherent in Scripture as well as in Catholic social teaching. (2) To more effectively practice what the Church preaches to the poor regarding conservation and sustainability.	CRS employees are energized by the new building, and feel that they are more effectively practicing the values that CRS promotes, particularly with respect to conservation and the importance of the sustainability of the earth's resources.
(1) To "practice what they preach," particularly the claim in the FCNL mission statement: "We seek an earth restored." (2) "Building green was a tangible way that FCNL could underscore its environmental lobbying efforts and engage its constituents on environmental issues."	There have been many tours in recent months, particularly by members of Congress, constituents, and other organizations considering green building projects.
Initiated by youth interest in tackling issues collectively. Subsequent discussions explored Islamic ethics of nonwasteful resource use. Stories from the Koran about using limited resources wisely served as inspiration to keep going. Religious and youth-interest elements helped drive project forward.	MKA received awards, including the Queen's Jubilee Award in 2003, and Best Environmental Demonstrator Building, bringing national attention, frequent visitors to see green features, and increased use of the space by community members.
To support Israel's quest for peace and security by promoting initiatives that decrease the world's dependence on foreign oil.	Still under construction.
(1) To more fully live out our Franciscan values, particularly caring for God's creation. (2) To create a more healthy living environment for the sisters, especially the infirm.	(1) Sisters are frequently asked to share their story about the building and the community. (2) Greater respiratory health among the sisters, especially the infirm. (3) The sisters have begun to "live green" by using environmentally friendly goods and reducing consumption, in keeping with Franciscan values of poverty and simplicity.
As reported in the *Elgin Daily Herald* on June 29, 2007, "Judson President Jerry Cain believes two primary motivators inspired the conception of the idea. 'We were primed to break out of the pack, and do something that said we are better than average,' explained Cain, 'There is also a theological reason. . . . We are responsible for everything God has given us.'"	(1) The new green academic center inspired Judson to initiate a campus-wide recycling program and energy-saving initiatives. The building has inspired students in architecture and environmental studies programs to incorporate green technology into their academic work. (2) Green building has inspired others in the city of Elgin to undertake green initiatives.
"As a university affiliated with the Lutheran tradition, our mission 'is to educate students for lives of thoughtful inquiry, service, leadership, and care—for other persons, for their communities and for the earth.' Since caring for the earth is a part of that mission, we decided that somehow we have to do that in the daily operations of our campus. Faith was not a primary motivating factor. PLU was motivated to build green because of growing environmental consciousness in the northwest U.S. Many other universities in the area have also built green buildings, and so to some extent, PLU felt a need to 'keep up with the Joneses' and to 'be a good neighbor.'"	Since environmental concerns and programs on campus predate the construction of the Morken Center, it is difficult to assess how much the green building has increased those efforts. However, following the completion of the Morken Center, 30% of all food purchased for student meals is organic or locally grown. Students now also discard food scraps for composting, and 60% of all campus waste is recycled. The presence of the green building on campus is also a major recruiting tool for new students.
The institution's motivation might be best summed up in the statement, "The University of Scranton's Catholic and Jesuit identity inspires and informs our attention to sustainable development. As a Jesuit university we are called by Saint Ignatius to 'seek God in all things.' As a Catholic university we manifest a deep concern for social justice and equality."	Ultimate goal is to educate students regarding sustainability. To achieve this, the university is providing faculty with incentives and training on ways to infuse environmental sustainability issues into the curriculum. As of May 2007, 50 of the 260 full-time faculty have done so.

TABLE 3.1 *(CONTINUED)*

Project	Religious affiliation	Green standard; year achieved or anticipated	Location	Data source
Sidwell Friends Middle School	Quaker	LEED Platinum, 2006	Washington, D.C.	Michael Saxenian, assistant head-of-school and CFO
Student Center, Seattle University	Catholic	LEED certified, 2002	Seattle	Michel George, associate vice president for facility services; Karen Price, campus sustainability manager
Vincent and Helen Bunker Interpretive Center, Calvin College	Reformed	LEED Gold, 2004	Grand Rapids, Michigan	Frank Gorman, college architect

HOUSES OF WORSHIP

Saddleback Church	Baptist	LEED equivalent, 2008	Lake Forest, California	Karen Kelly, director of campus development
All Saints Parish	Episcopal	Energy Star, 1999	Brookline, Massachusetts	Tom Nutt-Powell, property committee
Unitarian Universalist Church of Fresno	Unitarian	LEED Silver, 2007	Fresno, California	George Burman, project manager
Deer Park Buddhist Center	Buddhist	LEED equivalent, 2007	Oregon, Wisconsin	Ani Jampa
Jewish Reconstructionist Congregation	Jewish	LEED Gold, 2008 (anticipated)	Evanston, Illinois	Rabbi Brant Rosen and Julie Dorfman, head of the environmental committee
Adat Shalom synagogue	Jewish	Energy Star, 2001	Bethesda, Maryland	Rabbi Fred Scherlinder Dobb

TABLE 3.1 *(CONTINUED)*

Motivation	Impact
According to Sidwell's Web site, "with the decision to construct a new Middle School, Sidwell Friends chose sustainable design as a logical expression of its values. We believe that a 'green building' provides an opportunity to achieve an outstanding level of integration between the curriculum, values and mission of the school."[1]	Building green "substantially altered our culture." Environmental stewardship is now seen as one of the pillars of the school philosophy, along with academic excellence and diversity. Impacts include educational enrichment, extensive media coverage, and inspiring other environmental projects, including student-initiated clubs around global warming, composting, and recycling.
While environmental concerns have a long history at Seattle University, sustainability has become an increasingly important issue within the university and the Society of Jesus Oregon Province (the Jesuits), which is the religious order that sponsors the institution. "With our new master plan," notes George, "Seattle University is striving to become a leader in sustainability, both among Jesuit and non-Jesuit universities."	Seattle University's 20-year Facilities Master Plan was updated for 2006–2026. The plan is a guide and resource to meet the facilities needs of the campus. One of the plan's six goals is to "incorporate the principles of sustainable design in all aspects of site and building design, construction, maintenance and operation."
(1) To be good stewards of the resources that God has given us. (2) To construct a building that complements the 100-acre ecosystem preserve upon which it is located.	Has led to the infusion of sustainability principles and the biblical concept of stewardship of the earth within various departmental curriculums.
Primary motivation was "based on God's desire that we exercise stewardship of all our resources—both environmental and financial. For the environmental resources, we are looking for ways we can contribute to the world community by exercising green building options that make sense to us financially. The financial part of the equation is necessary because we are also required to be good stewards of the money given to us by our congregation."	Although the new building for student ministry is still under construction (70% complete as of February 2008), the church staff is beginning to become more environmentally conscious. For example, Styrofoam cups have been replaced by ceramic mugs, and recycling initiatives have received increasing emphasis. Kelly also stated, "within the church community, people are surprised by the number of members on the bandwagon behind this project and other environmental matters."
Financial savings were the initial motivation for the efficiency retrofit. Final decision to proceed was significantly influenced by the strong environmental stewardship motivation of the Property Committee, which served to "trump" concern about first costs.	Not available
Faith motivation was strongest factor. The Seventh Principle of Unitarian Universalism states that we have "respect for the interdependent web of all existence of which we are a part."	"Our members are proud of green building. The congregation voted to become a Green Sanctuary church, which will result in a 12-point action plan covering all aspects of our church's operation and programs."
Primary motivation is based upon Buddhist principles of nonviolence to others, including the environment. Also, as a sacred space for meditation and teaching, Buddhist temples are themselves symbolic and convey to others principles for living.	"People are impressed by the Tibetan detail and the highly symbolic building" according to the Dalai Lama, who visited for the opening of the temple.
Began as a grassroots interest in the connection between environmental issues and the Jewish faith.	According to Rabbi Brant Rosen, building the first certified "green" synagogue in the world "galvanized the congregation and created a sense of excitement that they were putting into practice an important aspect of their Jewish tradition."
The congregation's consciousness that environmental sustainability is an ethical and moral imperative that is intimately connected to the Jewish tradition and cannot be ignored.	The synagogue has further deepened the environmental consciousness of the congregation's 480 families, totaling 1,500 members, who are now striving to live greener at home and where they work.

1 See www.sidwell.edu/

3.3. Motivation

The primary similarity among most faith groups surveyed is the perception of congruence between the decision to build green and their faith tradition—in particular, the religious and moral or ethical imperative to be good stewards of the planet. An example of this perception is the Felician Sisters convent and school in Coraopolis, Pennsylvania, just outside of Pittsburgh. For these members of an international congregation of Franciscans (founded in Poland in 1755), the decision to build green was elementary. Franciscans strive to emulate Saint Francis of Assisi, who is often regarded as a patron saint of ecology and the environment. "As Franciscans, one of our primary religious values is 'Caring for Creation' by being responsible stewards of the resources given to us by God's loving providence," said Sister Mary Christopher Moore, one of the community's members who played a primary role in the green renovation of the convent and school.[16]

Our Lady of the Sacred Heart High School, which had been built in 1932, was deteriorating; the sisters either had to renovate or to construct a new building. The sisters approached the decision to build green gradually, beginning with curiosity about nearby green buildings, such as the one that had been built by the Greater Pittsburgh Community Food Bank.[17] Sister Moore noted that the more the community learned about green buildings, the more they began to realize that renovating their building to be LEED certified had a moral dimension:

> Building green was the morally right thing to do, because it not only considered the sustainability of our community but also the sustainability of our world. And in a community of sisters whose future is uncertain due to declining vocations to religious life, sustainability has been an important issue to us for some time. Building green simply expanded our conversations on sustainability into a new and meaningful dimension.

The sisters chose to preserve the structure (and the many memories contained therein) and incorporate environmentally friendly ideas into the renovation. In addition to energy-saving features, the renovation included refinishing and reusing more than 300 original hardwood doors and transoms. Likewise, over an acre of the original hardwood flooring and more than one mile of trim was removed, refinished, and reinstalled in the new structure. The underlayment of paving on the site included more than 275,000 pounds of roofing materials salvaged from the original building.[18] While some recycling resulted in financial savings, sometimes it was more costly than conventional construction. "This was a challenge for us as a community," said Sister Moore, "because we also wanted to be good stewards of our financial resources, many coming in the form of donations."

Like the Felician Sisters, various Jewish congregations have been interested in making a connection between their Jewish identity and the environment. Two such groups are the Adat Shalom Reconstructionist Congregation in Bethesda, Maryland, and the Jewish Reconstructionist Congregation (JRC) in Evanston, Illinois. Adat Shalom completed an environmentally friendly synagogue in 2001 and was given the Energy Star for Congregations award by the EPA the following year. The JRC's synagogue was completed in 2008 as a LEED Platinum building. In both cases, the decision to build green began as a grassroots effort, driven by interest from within the community and guided by leadership of the rabbis. In the case of the JRC, the initiative began when several members of the social action committee became passionate about environmental sustainability. Around the same time, the JRC's synagogue building was beginning to experience some major repair and maintenance problems, necessitating renovation or new construction.

The idea of building a green synagogue emerged as the congregation became more aware of the strong connection between Judaism and the environment. "Judaism and the environment have a long tradition, going back to teachings that during times of war, the Israelites were not to destroy trees, particularly fruit-bearing ones," notes Rabbi Brant Rosen, of the JRC; "this evolved into other prohibitions against squandering, wasting, and destroying the goods of the earth." These environmental concerns are inherent in the Jewish principles of bal tashchit, meaning "do not destroy or waste," and tikkun olam, which means "healing the earth." Wanting to learn more, the congregation asked Rabbi Rosen to speak to them in greater detail about the association between Judaism and the environment. In June of 2007, a statement made by Rabbi Rosen was quoted in the Chicago Jewish News:

> From the beginning, I felt very strongly about educating the congregation that this effort is grounded in our spiritual values as Jews. Environmentalism is not just a political issue, not just a bandwagon to jump on. It goes back to the Torah, a value we've inherited in our own spiritual tradition . . . energy efficiency, not destroying natural resources. The world does not belong to us. . . . We're reminded repeatedly of that in the Torah.[19]

Understanding this connection was crucial to achieving support from those members who were initially ambivalent about or against the green building project. "It helped us to see how caring for the earth was a tangible way to live out our Jewish beliefs," notes Julie Dorfman, head of the congregation's environmental committee. Making this connection was also crucial for fund-raising purposes, because financing the synagogue required contributions from all of the members. "If even a few had been adamantly opposed to building green," says Dorfman, "it would have been problematic."

The new green temple at Deer Park Buddhist Center, in Oregon, Wisconsin, was born out of a need to have a building that "reflects the concepts of Buddhism," which dictate that the "building itself should have a minimal impact on the environment." This desire to reduce the impact of the building included being "friendly to those in the building," through low-emitting paints and flooring. According to Ani Jampa, "[It was important] for the air to be as pure as possible since one of the primary functions of the temple is as a place to meditate. The building itself serves as a source of teaching. It's a very powerful space, very symbolic and lofty."

In 2005, the Friends Committee on National Legislation (FCNL) renovated its national headquarters to become the first LEED building on Capitol Hill. Founded in 1943 by members of the Religious Society of Friends (Quakers), the FCNL is now the largest peace-lobbying group in Washington, D.C. The FCNL played an important role in lobbying for the creation of the Peace Corps and the Arms Control and Disarmament Agency, and for the passage of the Civil Rights Act. Today, the FCNL is advocating political solutions to global warming, in an effort to fulfill the last line of its succinct mission statement: "We seek an earth restored." Located just a few blocks from the U.S. Capitol, the FCNL frequently lobbies Congress on global climate change as well as on alternative energy, dependence on foreign oil, and other environmental issues. "Building green was a tangible way that FCNL could underscore its lobbying efforts and engage its constituents on environmental issues," states Maureen Brookes, communications program assistant. "It was also a good means of practicing what we preach."

The FCNL's decision to undertake a green renovation came after careful consideration of the organization's core values and evaluation of many options. Renovation was chosen over new construction because of the FCNL's prime location near Congress and other constituents it seeks to engage and persuade. The decision to build green was based on various principles and beliefs long embraced by Quakers—particularly simplicity, equity, and stewardship. These values were echoed by Joe Volk, FCNL executive secretary, in a press release dated September 4, 2007:

> LEED certification affirms our General Committee's decision in the late 1990s to rebuild our two aging row houses on Capitol Hill, using green technology that allows us to walk more lightly on the earth and that provides a model for energy conservation that others might follow. Buildings in the United States account for nearly 50% of U.S. energy consumption and more than 40% of carbon dioxide emissions. We are working with Congress to promote green building design.[20]

Catholic Relief Services (CRS) has also sought LEED certification for its new world headquarters in downtown Baltimore, both as a way of putting its own words and values into practice and as a means of educating and influencing others. Founded in 1943 by the USCCB to serve World War II survivors in Europe, CRS works to assist the world's poor and disadvantaged by promoting the development of people everywhere, regardless of religion, race, or ethnicity. CRS achieves its goals by leveraging the teachings of the Catholic Scripture and by promoting the principles of Catholic social teaching. One such principle that is congruent with green building is stewardship.

For CRS, stewardship means that "there is an inherent integrity to all of creation and it requires careful stewardship of all our resources, ensuring that we use and distribute them justly and equitably—as well as planning for future generations."[21] Three resources that CRS encourages people to steward wisely are water, forests, and energy. In its effort to practice what it preaches, Dave Piraino, executive vice president of human resources, notes that CRS decided to incorporate similar principles of stewardship in the construction of its new headquarters. For example, Piraino notes that just as CRS encourages many people in developing countries to capture rainwater and to use it for multiple purposes (e.g., shower water can be reused to flush toilets), the organization decided to install waterless urinals and low-flush toilets in its restrooms. Similarly, in Haiti, where CRS has been concerned about soil erosion caused by deforestation, its new building was constructed with certified wood from forests that are managed to prevent overharvesting.

A second guiding principle for green building at CRS is the notion of the common good, which emphasizes the importance of solidarity among all people. According to Catholic social teaching, all of humanity is essentially one family—brothers and sisters. In essence, by making inefficient use of resources such as energy and water and contributing to the burdens of pollution and environmental degradation that fall disproportionately on the poor, conventional building practices violate the principle of the common good.

At Calvin College, a Christian institution in Grand Rapids, Michigan, the decision to construct the Vincent and Helen Bunker Interpretive Center as a LEED Gold building emerged largely from two sources: the presence of a 100-acre ecosystem preserve on the campus (70 acres of which are owned by Calvin College), and the grassroots commitment of faculty and students to environmental concerns. The establishment of the preserve began in 1964, when the college first acquired 25 acres of wetlands, mixed-hardwood forests, agricultural fields, and a horse farm. Fourteen years later, in 1978, a study committee from the Calvin Center for Christian Scholarship advised that, as a fitting expression of environmental stewardship, those acres should be set aside as a nature preserve.

This concern for the environment and Calvin College's heritage as a Christian institution also led to the infusion of principles of sustainability and the biblical concept of stewardship of the earth within various departmental curricula. For example, the Department of Geology, Geography, and Environmental Studies explicitly expresses a Christian commitment to caring for God's creation. One of its Web pages quotes Psalm 24:1 "The earth is the Lord's, and everything in it, the world, and all who live in it." On the same Web page, the department also states: "In the Department of Geology, Geography and Environmental

Studies, faculty, students and staff: analyze Earth's environmental systems and foster the commitment to serve God in their care and preservation."[22]

The biology department and the engineering department have also integrated academics, the environment, and Christian values. The biology department's Web page states that "studying biology at Calvin equips students to assume their roles and responsibilities as servants and stewards of God's creation."[23] Similarly, the engineering department, which offers a specialization in civil and environmental engineering, states that "using science and technology creatively to serve society points to the moral responsibility of the engineer."[24]

According to Frank Gorman, college architect and a member of the Environmental Stewardship Committee, this focus on the integration of Christian principles and academics is at the heart of Calvin College's Statement on Sustainability, which declares, "Our purpose is to infuse Calvin's vigorous liberal arts education with thoughtful, Biblically based practical guidelines that lay a foundation for living in a way that honors the Creator and his beloved creation."[25] Building the Vincent and Helen Bunker Interpretive Center in 2004 was simply another means to achieve this goal.

Sidwell Friends Middle School found a similar congruence between green building and institutional values. As stated on the Sidwell Web site: "With the decision to construct a new Middle School, Sidwell Friends chose sustainable design as a logical expression of its values. We believe that a 'green building' provides an opportunity to achieve an outstanding level of integration between the curriculum, values and mission of the school."[26]

3.4. Impact of Green Buildings in Faith Communities

Collectively, faith-based organizations have the potential to make a major environmental impact. In the United States alone, 63% of Americans belong to religious organizations, and over 40% of all households make charitable contributions to religious groups.[27] As with other building types, the design and operation of religious buildings has direct impacts on occupants and organizations. Green design typically reduces energy and water costs and improves occupant comfort. Reported energy and water savings in eight of the buildings surveyed indicate that over 20 years, the discounted present value of savings is $4/sf to $9/sf. In general, academic facilities, offices, and community centers have higher energy and water use than places of worship because of more continuous occupancy, and thus higher potential savings (though some places of worship may gain significant savings by conditioning building spaces only during use).

In addition to such financial savings and environmental impacts, our research indicates that building green positively contributes to these faith-based communities in a number of physical and spiritual ways. One reported physical benefit is improved health, achieved through better air quality. At the recently greened Felician Sisters Convent, for example, one elderly sister whose respiratory problems required her to sleep sitting upright, while using an oxygen inhaler, can now sleep lying down, without the inhaler. The teachers in the Felician Sisters school also report that the students are more alert and attentive in class, which they attribute to the improved air quality. Finally, the sisters note a difficult-to-measure but significant attitudinal improvement in many of the members of their community. Similarly, occupants of the new green FCNL office noted that the indoor environment of the new building simply makes it a "great place to work."

A second commonly reported impact from building green is a greater sense of empowerment in relation to the organization and its mission. At the Unitarian Universalist Church, for example, in Fresno, California, the success of the green building construction motivated the congregation to seek Green Sanctuary status, which involves the adoption of a 12-point action plan addressing all aspects of church operations and programs, and encourages energy savings and environmental practices in the homes of church members. Similarly, many employees at CRS recognize that the investment in their green building is motivating them to live out the organization's mission in new and tangible ways; according to Dave Piraino, the new green building "has had a huge impact upon people. They are energized that they too are practicing many of the values that CRS as an organization promotes around the globe." For example, instead of using energy-consuming elevators, many employees are using the building's staircase, which is centrally located to encourage use. Some CRS employees also bike to work, a practice made easier by the inclusion of shower and locker facilities and a safe storage area for bicycles. Many employees are particularly proud of the building because they participated in one of the 13 committees that CRS formed to help its architects design and develop the structure.

Building green has inspired additional environmentally conscious decisions and initiatives, particularly within the faith-based colleges and universities in our study. At Judson University, an evangelical Christian university in Elgin, Illinois, architecture students have been inspired to incorporate energy-efficient and environmentally friendly principles in their curricular projects. Student interest in environmental issues also led the university to develop an environmental studies program that began in the fall of 2006.

At Pacific Lutheran University, in Tacoma, Washington, recycling has been a way of life for many years; in 2006, 60% of campus waste was recycled. Building on this tradition, the 2006 completion of the green Morken Center for Learning and Technology coincided with a student-led initiative to begin composting the food scraps from meals in the cafeteria.[28]

The University of Scranton, founded and sponsored by the Jesuits, is seeking to become a better steward of God's creation through initiatives that foster sustainability and address global warming. For this reason, the university designed its new campus center—completed in January 2008—to be a LEED-certified building. This building is part of a wider initiative to incorporate sustainability into the university's academic programs. According to Springs Steele, associate provost for academic affairs, in 2005 the university began hosting a week-long seminar each May to encourage and help faculty members develop ways to integrate sustainability principles into their courses. As of May 2007, faculty members had revised 33 courses in 15 departments. Similarly, at Furman University, in Greenville, South Carolina, construction of several green buildings has supported further integration of sustainability concepts into curricular and long-range strategic planning. Such examples reveal that green buildings often inspire students, faculty, and staff to consider other initiatives that address environmental issues and concerns.

Faith-based organizations provide direct services to those in need but also serve as guides for the larger communities in which they reside. For instance, officials at Judson University believe that their green building inspired the city of Elgin to launch a number of green initiatives, including the purchase of hybrid cars for city employees, the construction of an eco-friendly fire station, and a program to collect rainwater to water plants and grass. Numerous members of Congress have toured the FCNL headquarters to learn more about green building. The building also inspired a visit by the Architect of the Capitol, who was seeking ideas for greening the Capitol Building. The FCNL hopes that such visits will translate into federal and state policies to help curb global warming and support greater protection of the environment. The Felician Sisters sometimes travel to other organizations to speak about their green building experience; they see such presentations as opportunities to "evangelize" to others about their biblical faith and Franciscan values.

Perhaps the most significant and immediate impact of green building is its ability to galvanize and energize community life around the process of green building. In a number of the organizations surveyed, the process of planning, approving, building, using, and thinking about the green goals and green strategies was an enriching and animating element in the life of the community. Green building efforts fostered intellectual and social engagement in the planning and building process, enhancement of the physical space for worship, and cross-generational educational enrichment.

Green renovations significantly reenergized community life at the award-winning Muslim Khatri Association in Leicester, England. The process of planning the renovations and becoming educated as a community on green features, such as solar panels and energy efficiency, brought together younger and older generations of the community. While interest and energy from younger community members inspired and drove initial inquiries into green design, the older community members were able to provide guidance and education on Islamic values and stories dealing with issues of sustainability. Use of the center increased by more than tenfold after the renovations, from 150 users per week to 2,000 users per week.

3.5. Financial Stewardship

Our study indicates that financial considerations were not the primary motivation for deciding to build green. However, the cost-effectiveness of green building strategies helped community leaders secure broad buy-in for green initiatives. Because houses of worship in particular are not big users of energy (largely because their occupancy level is typically lower than offices and homes), payback periods for initial investment in green measures can be longer for houses of worship than for other green building types—often 15 to 20 years from energy savings alone. Faith communities are often funded through voluntary member donations, and our respondents repeatedly mentioned that being responsible caretakers of financial resources was a priority. Consequently, many of the groups surveyed indicated that the potential cost savings associated with green renovations led to discussions about the ethical nature of cost trade-offs, and the appropriate reinvestment of potential cost savings in other charitable activities. In some cases, green goals helped build necessary community support for new buildings, as noted by Julie Dorfman, of the Jewish Reconstructionist Center: "The grassroots process and decision to build green galvanized the congregation's commitment to fund the project; that would not have happened if conventional construction was used and they were not so involved in the decision-making process."

The cost-effectiveness of green building among a wide range of faith groups represented in this study had a significant impact on broadening the potential spiritual and material contributions each faith community could make. The affordability of green building ultimately enabled these faith communities to express their core spiritual tenets and moral vision for a better world through creative, dynamic, and inclusive processes.

3.6. Conclusion

Compelling moral forces are moving people of faith to care for the environment, and these forces spring from the very heart of their traditions: growing reverence for the earth, acknowledgment and repentance for widespread assaults on the environment, daily enactments of sacred scriptural mandates to care for the planet, the pursuit of justice and equality, and acknowledgment of a responsibility to provide for future generations. These green building initiatives are sustained by sacred writings, by informed leadership, by a spirit of community, and by the deeply rooted religious and moral perspectives that give the actions of faith communities renewed meaning.

PART IV

GREEN DESIGN, CLIMATE CHANGE, AND THE ECONOMY: POTENTIAL IMPACTS IN THE UNITED STATES

Green design is less expensive and more cost-effective than generally realized. Moreover, because buildings account for 45% of U.S. energy consumption, the built environment offers an opportunity for tremendous positive economic and environmental impact—including on climate change. A shift to green design would increase investment in areas such as insulation, renewable energy, and recycling, while cutting energy use and creating net new jobs. If green design were scaled up nationally, the employment, financial, and environmental benefits would be large. The cost-benefit analyses in this book provide a basis for calculating the financial and CO_2 impact of a robust national transition to green design.

In its 2009 budget, the Obama administration committed to cutting U.S. greenhouse gas emissions 14% by 2020 from 2005 levels; and by 80% by 2050. This goal represents a monumental shift in policy and an enormous financial, political, and technical challenge. The good news is that green buildings and energy efficiency have the potential to cost-effectively drive deep reductions in energy costs and in CO_2 emissions. Such a strategy would also create large economic and social benefits.

To calculate potential reductions in CO_2 emissions from the building sector, we modeled the savings that could be achieved under two different scenarios—Business as Usual (BAU) and Building Green (Green)—through 2030.[1] In the BAU scenario, gains in energy efficiency and green building, including renewable energy, partially offset the growth in energy use caused by continued expansion of U.S. building stock. This BAU scenario is based on, though greener, than Energy Information Agency [EIA] projections. In the Green scenario, green design and construction become the industry standard, and green buildings (including more rapid retrofits of existing buildings) drive relatively rapid and sustained increases in energy efficiency and renewable energy. Thus, despite steady growth in the building stock, CO_2 from U.S. buildings drops to 14% below

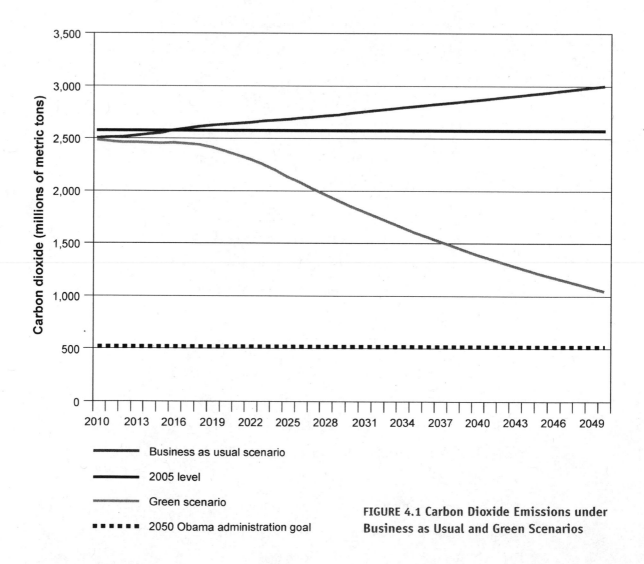

FIGURE 4.1 Carbon Dioxide Emissions under Business as Usual and Green Scenarios

2005 levels by 2025 and to almost 60% below 2005 levels by 2050 (see figure 4.1, which shows the trends in terms of millions of metric tons).

A sustained national commitment to green design would create tremendous financial, social, and environmental benefits. As this book has shown, the costs of building green are far outweighed by the financial benefits, which include reduced energy and water costs, enhanced health and productivity, and broad societal benefits. Applying the cost-benefit findings from our study data set to the two scenarios shows that, compared with the BAU scenario, the Green scenario creates $650 billion more in net financial benefits (see figure 4.2). Why? Green buildings generate financial benefits that are five to ten times as large as their green cost premium. Additional benefits that are not included in this analysis include lowered dependence on energy imports, increased employment, and increased economic competitiveness.

As discussed in section 1.2, materials use in building construction such

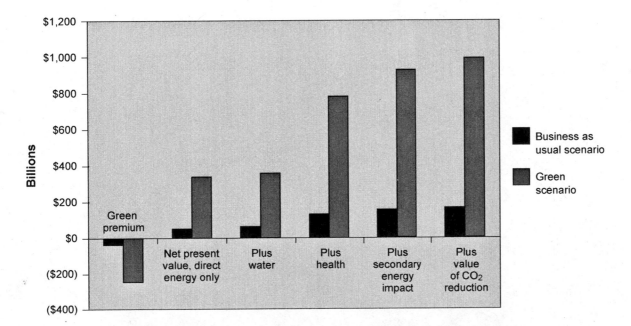

FIGURE 4.2 Net Present Value of Business as Usual versus Green scenarios
Note: The values shown for the columns labeled "Plus Water," "Plus Health," "Plus Societal Energy," and "Plus Societal Carbon Dioxide" are cumulative.

as concrete, steel and drywall are large industrial users of energy. The production of cement alone is estimated to cause about 2% of US CO_2 emissions and between 5% and 8% of global CO_2 emissions. Green buildings reduce energy and CO_2 by using recycled, low CO_2 and locally sourced materials. Similarly, the focus on waste reduction and a very high rate of waste diversion and recycling in green buildings also cuts energy use and CO_2 and is very labor intensive compared with sending waste to dumps. These CO_2 reduction and employment creation benefits and the CO_2 reduction from waste reduction are not included here.

Figure 4.2 assumes a CO_2 price of \$15 to \$20 per ton. Most analysts project that the price of CO_2 will be higher, in which case table 4.2 would show larger national financial benefits from greening. The societal *cost* of CO_2 due to its role in driving global warming is far larger than the price. If societal costs of global warming were included, the present value from a national shift to green design would likely be many trillions of dollars.

4.1. Energy Consumption

We modeled energy savings to 2030, then projected the trends an additional 20 years, to 2050.[2] Both scenarios are based on the 2009 annual projections for building sector energy consumption and carbon emissions made by the EIA.[3] The EIA's projections are based on a model that tracks U.S. primary energy consumption and energy-related CO_2 emissions for the residential, commercial/institutional, industrial, and transportation sectors by fuel source, and makes projections from the present day to 2030. Both scenarios use EIA's projections for new building construction and building demolition, and both define the energy efficiency of conventional and green new construction and of comprehensive conventional and comprehensive green retrofits in the same way: in comparison to the energy consumption of an average building in 2010.[4] Both scenarios project steady improvements in energy efficiency for these four building types,[5] and the energy performance for each building type is the same in both the BAU and Green scenarios (see figure 4.3).

In figure 4.3, 2010 vintage building energy consumption declines over time as new more efficient appliances and energy using devices replace less efficient ones. New non-green construction and efficiency retrofits show similar, gradual improvements. In figure 4.3, green new construction and green retrofits start at the level of energy use documented earlier in this book and are projected to decrease relatively rapidly, reflecting adoption of new and more effective energy efficiency and renewable energy technologies and systems.

There are only two initial differences between the scenarios:

- Whether new green construction and green retrofits remain a niche (albeit a substantial one), or become the norm
- How frequently existing buildings are comprehensively retrofitted.

The BAU scenario forecasts more rapid growth than EIA projections in market adoption of green building practices for new construction and comprehensive retrofits. The percentage of new green buildings constructed each year grows fivefold, from a 2010 base of 5% to 25% of new construction.[6] Similarly, the percentage of green retrofits increases from an estimated base of 0.25% to almost 5% by 2030.[7] Thus, as shown

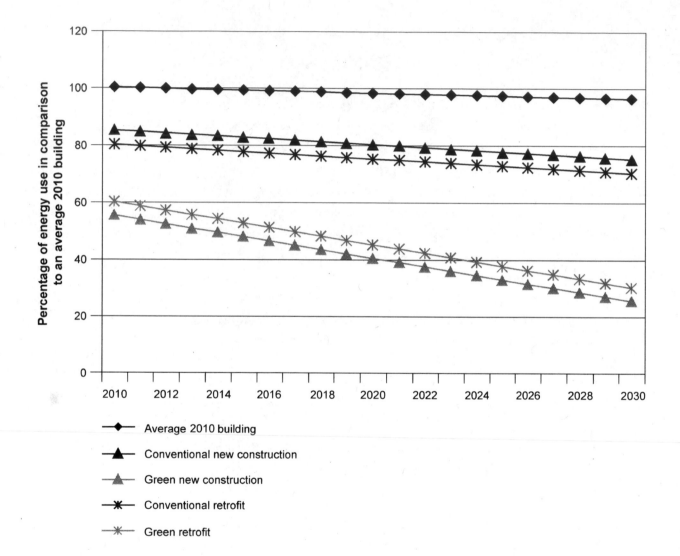

FIGURE 4.3 Energy Use in New Buildings or Retrofits versus Average 2010 Building Stock

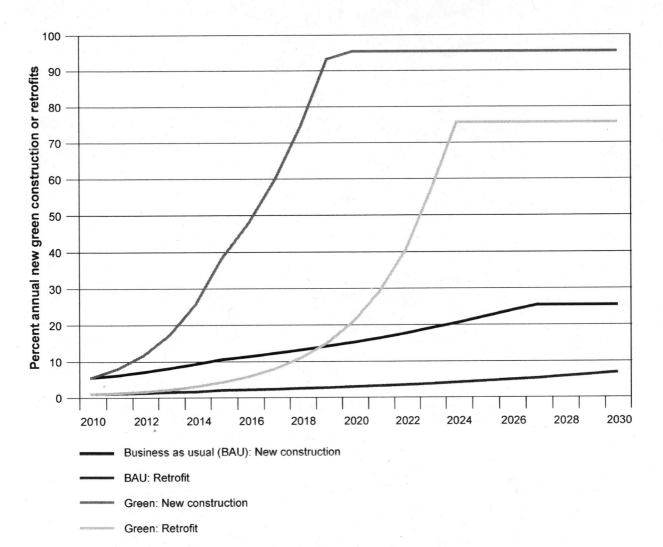

FIGURE 4.4 Green Building Growth: Business as Usual and Green Scenarios

in figure 4.4, while the green building market grows substantially over 20 years, it ultimately becomes only a large niche—not the design norm.

In the Green scenario, green new construction and retrofits start at the same base percentages as the BAU scenario, but then grow more rapidly and become the design norm. The portion of new green buildings increases by 50% per year between 2010 and 2015 (slower than the rate of growth in LEED new construction from 2003 to 2009),[8] and slows to 25% per year between 2016 and 2030. The percentage of comprehensive green retrofits grows at 50% per year until 2015, then slows to 40% between 2016 and 2030.[9] As illustrated in figure 4.4, green design ultimately becomes standard practice for 95% of the new construction market by 2020,[10] and for 75% of retrofits by 2030[11]— reflecting the likelihood that some types of buildings will remain unlikely candidates for

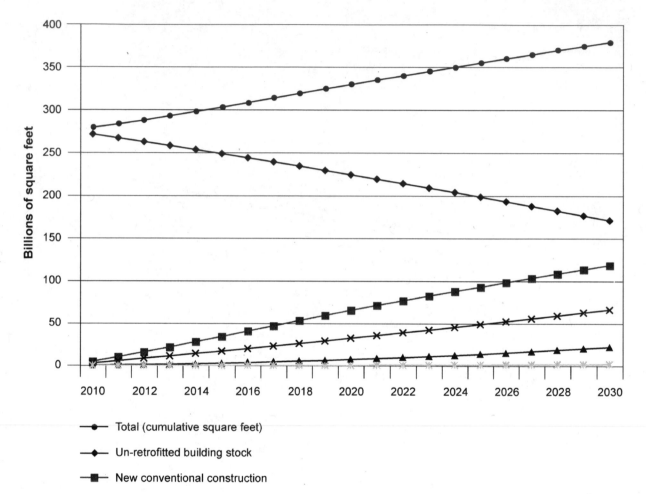

FIGURE 4.5 Cumulative Building Floor Area by Type: Business as Usual Scenario

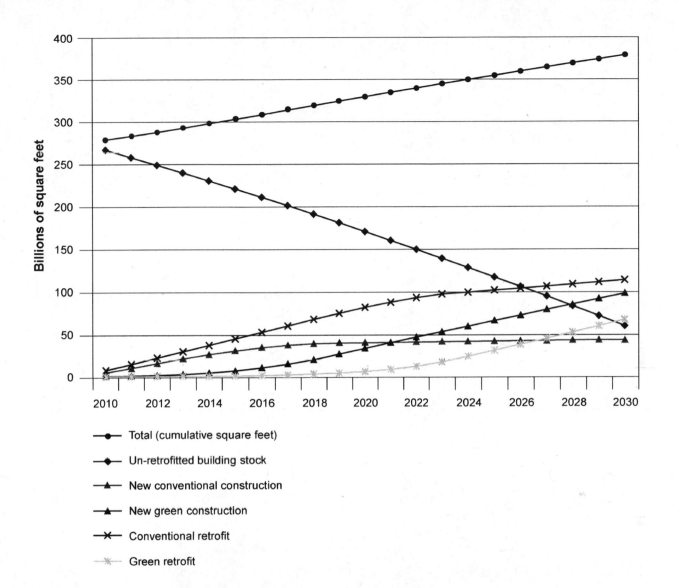

FIGURE 4.6 Cumulative Building Floor Area by Type: Green Scenario

greening. The lag in greening retrofits reflects the fact that current green penetration of the retrofit market is far lower than the penetration of the new construction market.

Buildings typically last 50 years or more. By contrast, an automobile fleet is typically replaced every 12 to 15 years. Not surprisingly, deep reductions in energy use in the building sector cannot be achieved quickly—and cannot be achieved by 2050 without a large increase in rate and comprehensiveness of retrofits of existing buildings. The Green scenario assumes that comprehensive retrofits, whether as energy-efficiency retrofits or as part of greening, occur more frequently than in the BAU scenario. In the BAU scenario, existing building stock is turned over—retrofitted or demolished—once

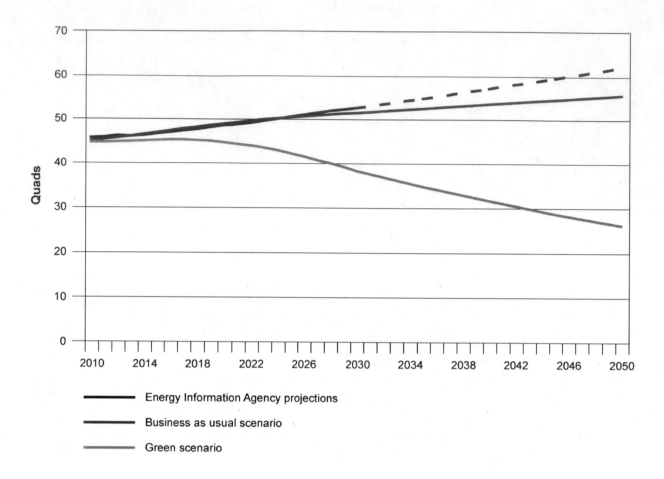

Energy Information Agency projections

Business as usual scenario

Green scenario

FIGURE 4.7 Building Energy Consumption: Energy Information Agency, Business as Usual, and Green Scenarios

every 45 years; in the Green scenario, turnover occurs once every 25 years, reflecting an increase in the frequency of greening or comprehensive energy-efficiency retrofits (see figures 4.5 and 4.6.)[12]

We are more optimistic than the Energy Information Administration (EIA) about the market's adoption of energy efficiency, renewable energy, green building practices, and whole-building green retrofits; thus, the BAU scenario anticipates a moderately more energy-efficient building sector, using more renewable energy, than EIA projections. Our Green scenario sees a more substantial and rapid decrease in overall building sector energy consumption from 2018 on (see figure 4.7). In 2030, the Green scenario uses 26% less nonrenewable energy than the BAU—and in 2050, 60% less nonrenewable energy.

4.2. Renewable Energy

In addition to being more energy-efficient than conventional buildings, green buildings are more likely to generate energy on site[13] or to purchase energy or renewable-energy certificates (RECs) from renewable energy sources.[14] On the basis of USGBC data, we estimate, conservatively, that new green buildings and green retrofits will in 2010 consume 12% and 2% of their energy, respectively, from off-site and on-site renewable sources. We estimate that the portion purchased from off-site sources will grow 15% a year, while on-site energy will grow 10% a year. This reflects declining costs of renewable energy (solar PV panel prices dropped by 30% between mid-2008 and mid-2009), increasing levels of greenness, and growing demand for low- and zero-net-energy buildings. We used

FIGURE 4.8 Annual Renewable Electricity Use: Business as Usual versus Green Scenario

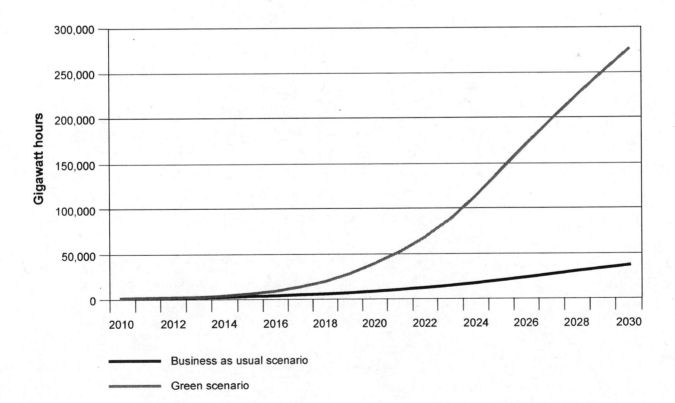

- Business as usual scenario
- Green scenario

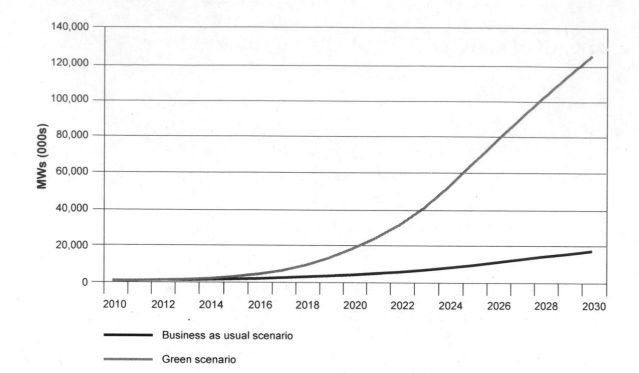

FIGURE 4.9 Cumulative Installed Renewable Capacity: Business as Usual versus Green Scenario

these projections as inputs to calculate how much renewable energy is consumed by green buildings each year in the BAU and Green scenarios.[15]

Given uncertainties about the impact of purchases of off-site renewable energy, we discounted off-site renewable electricity and Renewable Energy Credit purchases by 50%.[16] In the BAU scenario, 50,000 gigawatt hours (GWh) of green electricity are consumed by 2030; in the Green scenario, the figure is 280,000 GWh (see figure 4.8). We then calculated the generating capacity needed to produce this much green electricity.[17] In the BAU scenario, the use of renewable energy by green buildings grows steadily, requiring about 0.7 gigawatts (GW) of new capacity a year by 2020, and 1.5 GW a year by 2030, for a cumulative installation of 16 GW between 2010 and 2030 (see figure 4.9). In the Green scenario, greater adoption of green building design drives more rapid deployment of renewable energy: 5 GW of renewable capacity a year are added by 2020, and 12 GW a year by 2030, for a cumulative installation of 124 GW between 2010 and 2030 (see figure 4.9.)

The perhaps surprisingly rapid growth in renewable energy demand from green buildings in the Green scenario reflects three facts and one assumption. 1) renewable energy such as solar photovoltaics and wind generate electrcity, 2) buildings use about 75% of electricity, 3) green buildings are about 30 times as likely to deploy on-site renewable energy or buy offsite renewables (in the form of green power or renewable energy credits) compared with non-green buildings, and (the assumption) 4) a continued rapid growth in new green and retrofitted green buildings.

4.3. Carbon Dioxide Emissions

We calculated building CO_2 emissions in the BAU and Green scenarios by applying a building sector energy-to-CO_2 conversion factor (derived from EIA and DOE projections)[18] to the energy consumption projected in each scenario. To calculate total emissions in both scenarios, we then calculated and factored in the CO_2 emissions reductions achieved through the use of renewable electricity.[19]

In the BAU scenario, increases in the energy efficiency of conventional design and substantial penetration of green building design offset some, but not all, of the increases in building stock through 2050. In the Green scenario, emissions drop quickly. The continued rapid and sustained penetration of green design and increased efficiency results in a 31% reduction in CO_2 by 2030 and, compared with BAU, a 26% reduction when compared with 2005 emissions levels. By 2050, CO_2 emissions in the building sector are 65% lower than those in the BAU scenario and 57% lower than 2005 emissions levels.

Figure 4.10 breaks out the sources of CO_2 reduction in the Green scenario. The majority of the CO_2 reduction results from gains in energy efficiency, but the contribution from renewables is considerable as well. Because there is greater uncertainty about the impact of green building purchases of off-site renewable energy (purchased as green power or RECs) than about the impact of on-site renewable energy, the impact of off-site renewable electricity is discounted in this scenario by 50%.

Although it comes close to achieving the Obama goal, the Green scenario misses target reductions. This shortfall is largely due to continued growth in buildings square footage, and to the historically slow turnover in building construction and retrofits (in other words, buildings last a long time).

As discussed earlier in this book, green communities encourage walking and the use of public transportation, and can therefore have a substantial impact on transportation emissions by reducing vehicle miles traveled and enabling walking, cycling, and public transportation. Further, because green buildings are far more likely to generate renewable energy on site and to have load-shifting devices (such as ice storage), demand management, and smart-control technologies that enable low-cost load reduction or load shifting, they offer greater potential for shifting and shaping electrical loads to flatten load

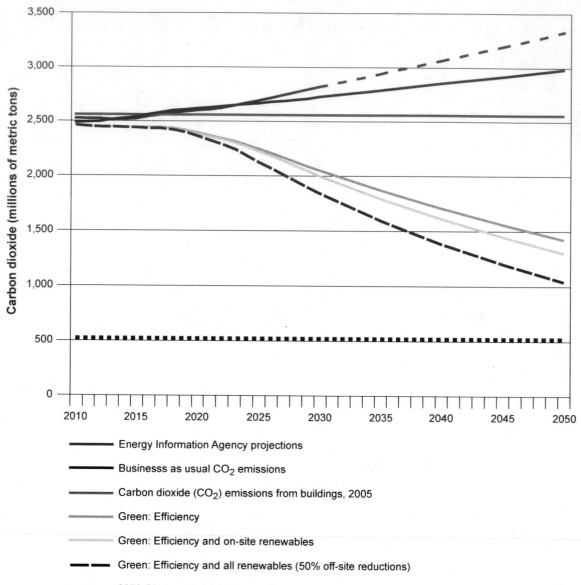

Legend:
- Energy Information Agency projections
- Businesss as usual CO₂ emissions
- Carbon dioxide (CO₂) emissions from buildings, 2005
- Green: Efficiency
- Green: Efficiency and on-site renewables
- Green: Efficiency and all renewables (50% off-site reductions)
- 2050 Obama administration goal

FIGURE 4.10 Building Sector Carbon Dioxide Emissions

demand and to enable expanded renewable-energy generation. Lower peak demand reduces strain on the electrical grid and cuts associated transmission and distribution losses, and load shifting enables the use of PV and wind energy.[20]

However, even these additional reductions would not enable the building sector to fully achieve the 2050 target. The building sector is projected to grow significantly (35%) between 2010 and 2030, making building sector emissions reductions an uphill battle. In order to meet projected U.S. population increases, building space is projected to increase 73% between 2010 and 2050.

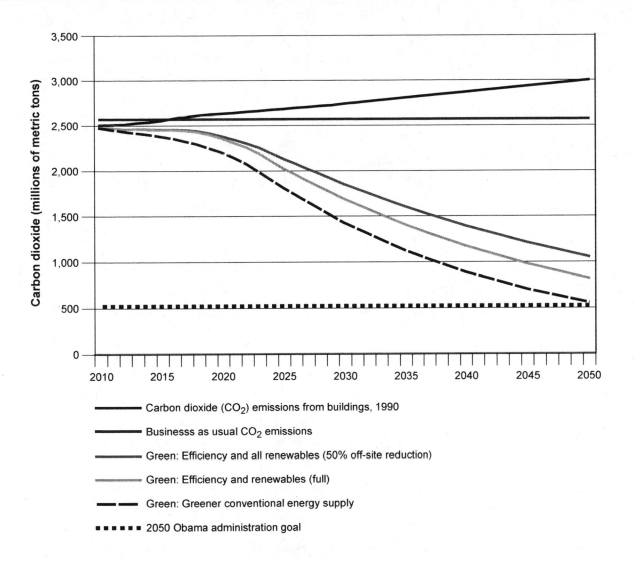

Legend:

— Carbon dioxide (CO_2) emissions from buildings, 1990

— Businesss as usual CO_2 emissions

— Green: Efficiency and all renewables (50% off-site reduction)

— Green: Efficiency and renewables (full)

- - - Green: Greener conventional energy supply

▪▪▪▪▪ 2050 Obama administration goal

Thus, it is essential to go beyond the energy efficiency and high use of renewables assumed earlier. What steps are available? First, we envision adoption of a LEED policy change to help ensure that purchase of off-site renewables results in net new green power purchases; such a change would help drive investment in and development of new renewable energy. Specifically, green power purchase would be restricted to more recent renewable energy projects, and projects would be required to sign longer-term contracts (e.g., for five years) for renewable electricity, as opposed to two-year purchases, which is the requirement as of mid 2009 (The LEED Energy and Environment Technical Advisory Group developed, and, in July 2009, unanimously endorsed these changes, which are expected to be submitted to ballot in late 2009 for adoption in beginning 2010). Such a policy would help to ensure that green power or renewable energy credits purchased drive the development of new renewable resources. The 50% discount applied to offsite re-

FIGURE 4.11 Impact of Additional Policy Measures on Emissions

newable energy purchases earlier could be eliminated; in effect, the full CO_2 impact of these green power/REC purchases would be recognized. Second, we assume the establishment of a federal CO_2 cap-and-trade program (setting a national cap on CO_2 emissions) that allows the price of CO_2 to directly drive investments in energy efficiency and renewable energy. To be effective, such a program would ensure that investors in energy efficiency and renewable energy (whether building owners, utilities or developers) would get the value of the resulting CO_2 reductions. If the value of the CO_2 resulting from efficiency and renewable investments accrues to the investors, it could offset 15% to 50% of the cost of clean energy investments, greatly increasing these investments. However, some cap-and-trade legislation would not do this and instead would, for example, have the value of the CO_2 created by building owners investing in efficiency default to the utility—even if the utility had nothing to do with the clean energy investments. Poor program design would sharply limit the CO_2 reduction impact of climate change legislation and unnecessarily prevent investors from receiving the value of the CO_2 reductions resulting from their investments in energy efficiency and renewables. Third, there must be substantial progress toward decarbonizing both the conventional electricity supply and the fuels consumed by buildings on site. Such a shift might result from a national renewable-portfolio standard designed to green the general electricity supply; it might also come from continued price and performance advances in the next generation of very low carbon energy sources, including wind, solar and biofuels, which would allow reductions in the carbon content of building electricity and heating fuel.

To show possible impacts, we have modeled the Green scenario assuming expanded and more rigorous LEED off-site renewables requirements, an effective cap-and-trade program, and a 15% decrease in the carbon intensity of the U.S. energy supply between 2010 and 2030 (see figure 4.11, which depicts trends in terms of millions of metric tons). With these changes, CO_2 emissions are reduced to almost 80% below 2005 levels.[21] Given that the turnover of buildings is far slower than that of transportation or industry, these findings indicate that an economy-wide CO_2 reduction of 80% by 2050 is feasible.

4.4. Financial Impact

To assess the financial impact of the BAU and Green scenarios we use findings and the data set,[22] developed earlier in this book, including energy and water savings,[23] health and productivity increases, and societal benefits such as lower energy prices from reductions in demand, and lower CO_2 emissions.[24] We then compared the NPV of green buildings and retrofits in the BAU scenario and in the Green scenario.

As indicated in figure 4.12, direct energy savings alone more than offsets the green building premium, providing net present value to society of over 300 billion dollars. Additional savings from water, health and secondary energy benefits increase the net financial benefits to over 800 billion dollars. As noted above, higher CO_2 prices, or an accounting of actual societal costs of CO_2 in driving global warming would greatly increase the financial benefits of the Green scenario. Figure 4.12 represents the financial impact of accelerated greening. Including benefits from accelerated efficiency retrofits would substantially increase the estimated financial benefits in the green scenario.

FIGURE 4.12 Net Present Value of Benefits: Business and Usual Scenario versus Green Scenario

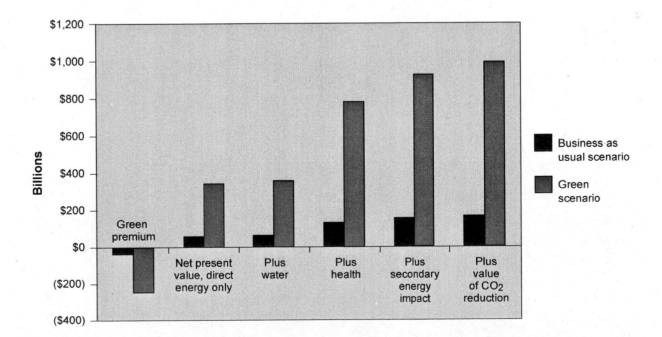

A national shift to green design strategies for how and where buildings are located, such as conservation development and transport-oriented development, would considerably extend the CO_2 reduction impact of green design. As detailed in part II, a shift to conservation development can both cut transport energy and CO_2 while reducing development costs by $12,000 per home site. As discussed in section 2.4, a shift toward residential design and zoning to increase walkability and access to public transport could drive substantial reduction in transport-related energy use and CO_2, as well as achieve large financial gains from improved health.

These shifts to green design would provide large financial savings, improve health and generally enhance building values. But these design approaches are today typically by exception rather than normal practice—and changing this will require widespread changes in legislation, codes and zoning practices.

The benefits of green building and green communities greatly outweigh the additional costs associated with high-performance design, materials, and technology. The financial benefits of a shift to green design probably offer the single largest opportunity to both strengthen the economy and address the monumental problem of global warming. The data and analysis in this book demonstrate that greening is highly cost-effective and has broad additional benefits for institutions ranging from schools and residential development to hospitals and places of worship.

Conclusion

Design consequences are at the center of many major environmental, economic, and societal challenges, including climate change, volatile and rising energy prices, energy security, high unemployment and socioeconomic inequity. A growing awareness of the health impacts of indoor environments has been accompanied by persuasive research establishing the link between community design and long term strength of communities as well as rates of physical activity—and, in turn, obesity. Greening buildings and development design is a very cost effective way to simultaneously address these challenges.

Some have argued that a rapid transition to a clean energy economy would be too expensive and cost America jobs. The data collected and analyzed in detail in this book demonstrate the contrary. A rapid transition to greening and a deep improvement in energy efficiency are highly cost effective and would create on the order of a trillion dollars in wealth, about $10,000 per family. A rapid transition to green design would be a large creator of employment, would strengthen US security and have large health benefits.

In the United States, buildings are responsible for almost half of energy use and CO_2 output. So, achieving the deep reductions in CO_2 emissions that scientists warn that we must achieve to avoid the most severe consequences of climate change will require deep and relatively rapid reduction in energy use in buildings. This can only happen with a huge increase in building energy efficiency and a rapid increase in the use of renewable energy. Greening buildings provides a very cost effective way to achieve both objectives.

New generations of higher performance and more cost effective energy efficiency and renewable energy technologies are being driven by expanded venture capital and Department of Energy funding. The combination of high efficiency and renewable energy make very low and even zero net energy buildings increasingly cost effective. Accelerating deployment of these technologies through rapid expansion in green design would position the United States to be the leading global supplier of clean energy technologies in the coming decades, with all that means for job creation and competitiveness.

State, local and federal policies will be critical to this transition. For example:

- A federal CO_2 cap and trade program could play a large role in driving increased investment in efficiency and greening if investors receive the value of the CO_2 reductions they create through their clean energy investments.
- Current building retrofits typically cut energy use by 15% to 25%, but once a building is retrofitted it is typically uneconomical to retrofit it again for years. To achieve deep improvements will require that these shallow "cream-skimming" retrofits generally be eliminated in favor of deep retrofits that achieve at least a 30% or 35% improvement in efficiency. Public funding or mandates for retrofits should generally require a minimum level of energy savings.
- Expanding financing for residential energy efficiency is a particularly large challenge. Public or utility funding can leverage eightfold if it is used to buy down the loan rate for financing home owner purchases of very efficient heating and air conditioning systems and windows. Such programs could scale quickly to millions of homes a year (see also www.islandpress.org/Kats).
- Current planning and zoning regulations and codes typically block or make slower and more expensive conservation development, mixed use and other green design approaches. Public de facto bias against green design is the norm despite the huge public cost savings offered by, for example, large reductions in waste water treatment facilities and other publicly funded infrastructure that green design delivers. These counterproductive public planning and zoning regulations and codes should be replaced with policies that recognize and support green design.

Design decisions are influenced not only by access to data-based cost and benefit information, but also by ethical and religious values. The green building movement has brought an awareness that decisions relating to buildings impact not only the aesthetics or function of physical structures, but also the long-term financial health of our households, businesses, and public institutions; the physical health of our families; the vitality of our cities and towns; the stability of our climate, energy, and water supplies; and the ecological health of the planet. A careful weighing of the costs and benefits demonstrates that the future will be far richer if we green our built world.

APPENDIX A
DATA-COLLECTION METHODOLOGY

We used a standard data-collection sheet to gather the data for part I of this book (the survey instrument is shown at the end of this appendix). Participation in the study was solicited from over 40 leading architecture, development, engineering, and consulting firms involved with green buildings. E-mail announcements soliciting examples of buildings to include in the study were also sent out to contact lists for the following: Building Green's High Performance Building Database; the New Buildings Institute's study of post-occupancy performance of LEED buildings;[1] LEED for New Development pilot-project applicants; the World Green Building Council; and the American Association of Sustainability in Higher Education newsletter.[2] The Green Guide for Health Care, the Austin Green Building Program, and the Massachusetts Technology Collaborative assisted by sharing lists of leading examples of green buildings participating in their programs. (See appendix B for a full list of participating firms and sources.)

Once points of contact for a firm or building were identified and had expressed interest in participating in the study, the standard data-collection spreadsheet was e-mailed to the source, along with a request for any relevant supplemental materials. Data sheets were returned to us with available data and supplemental project materials (e.g., case studies, LEED checklists, energy or water models, articles on projects). In many cases, further e-mail and phone contact was necessary to confirm and clarify reported green premiums, energy and water savings, and other data. In some cases, we extracted relevant data from supplemental materials, transferred it onto the data-collection sheet, then sent the data-collection sheet back to the data sources to confirm. Although we solicited a wide range of data, we assigned priority to the collection of data on green premiums, energy savings, and water savings.

Two previous studies, "Costs and Financial Benefits of Green Buildings" and "Greening America's Schools: Costs and Benefits," offer similar, broad-based financial evaluations of green buildings based on the development of new data and the synthesis of that data with existing research on green buildings.[3] A number of organizations have compiled public databases of case studies and performance information on green buildings, including the Building Green High Performance Building Database and the U.S. Green Building Council's database of certified and registered projects.[4]

From an initial list of over 300 potential additional projects to include in the data set, data on green premiums were available for 130 U.S. projects. The data set used in the study includes 40 buildings for which information on cost and performance had been collected through the two studies cited above. Information on these projects was confirmed through contact with the original data source or through corroboration from another online or published source.

To ensure that the costs associated with any benefit-producing strategies would be reflected in the analysis, the availability of data on the green premium was the threshold for inclusion in the final data set. Appendix C shows the primary data that was collected from the 155 buildings and used as the basis for this analysis. (Data on 15 non-U.S. examples of green buildings are also included in appendix C.)

The data-collection sheet was updated once in the course of data collection (in early 2007): specifically, certain fields were added or deleted to reflect the availability of information. For instance, the original sheet asked participants to compare energy use with the American Society of Heating, Refrigerating, and Air-Conditioning Engineers' (ASHRAE) 90.1 2004 baseline, which is used as the standard for LEED 2.2 energy-performance credits. It turned out, however, that many data sources had already conducted energy modeling or made measurements based on another benchmark (e.g., ASHRAE 90.1 1999, or California's Title 24), and it was not feasible to ask these sources to recalculate the savings based on the newer standard. Thus, a separate field was added to indicate the baseline for energy-use reductions. Similarly, multiple questions about health impacts were reduced to a single broad request for information, because detailed data were not available to answer the original and more focused survey questions.

DATA-COLLECTION SHEET

1—Project Info

a	Name of development	i	Number of buildings
b	City	j	LEED level or equivalent (silver, gold)
c	State	k	LEED points or equivalent
d	New construction or renovation?	l	Total building square footage (conditioned space)
e	Name and group (data source)	m	Total site area
f	E-mail	n	Number of occupants
g	Phone number	o	Year completed
h	Building type		

2—Costs

a	Cost of building ($/sf)
b	Green Premium (cost difference between actual building and the same building constructed using conventional building practices/without green features) ($/sf)
c	Green premium after incentives/grants
d	Cost of additional green features (e.g., on-site renewables, constructed wetlands, green roof, etc.) not included in "Green Premium" ($/sf)

3—Energy

a	Total building electricity use (kWh/sf/yr)
b	Total building gas use (Btu/sf/yr)
c	Total building energy cost ($/yr)
d	Reduction in energy use compared to conventional building (% below ASHRAE 90.1 2004 baseline or indicate baseline below)
e	Modeled or actual energy savings?
f	Baseline standard if not ASHRAE 90.1 2004
g	Energy savings per year ($/sf)
h	Total peak demand use (kW/sf conditioned space)
i	Reduction in peak energy use (%)
j	Is there a time of use tariff? If so, what's the peak rate? ($/kWh)
k	Green power purchase? How much for how long? On-site renewables?

4—Water

a	Total building water use (gal/sf/yr)
b	Reduction in water use compared to conventional building (% below EPAct 1992 baseline)
c	Reduction in waste-water compared to conventional building (% reduction)
d	Reduction in storm-water runoff from site (% reduction)

5—Health/Indoor Environmental Quality

a	Total IEQ points in LEED or equivalent
b	Describe any health impacts in building from green building features.

6—Materials

a	Long-term operations and maintenance costs or savings (e.g., reduced replacement or repair costs for longer-lasting flooring) ($/sf/yr)
b	Reduced or increased maintenance time due to more green materials or technologies (hrs/yr)
c	Construction and demolition waste diverted from disposal (tons)
d	C&D waste diversion (% of total waste)
e	% Recycled content of building materials
f	% Local building materials (w/in 500 miles)

7—Site

a	Development Density LEED points or equivalent
b	Alternative Transport LEED points or equivalent
c	Estimated % of occupants using public transportation
d	Estimated % of occupants walking or biking to work
e	Estimated % of occupants carpooling to work

8—Property

a	Change in property value due to greening
b	Change in rental rates/ occupancy rates/ speed of lease-up/sale

9—Other

a	Do any aspects of the project result in public infrastructure savings or costs compared to a conventional building (e.g., roads, sewer connections, storm-water drainage)?
b	Any other costs or benefits associated with greening? Insurance- or risk-related benefits? Accelerated permitting, subsidies, grants, or tax incentives because your building is green?

APPENDIX B
SOURCE LIST

(See also www.islandpress.org/Kats)

	Firm	Name
1	"Off-the-Shelf Ecology," *Building Design & Construction* (May 2001): 57-60	Langdon Wilson, C. C. Sullivan
2	"Summary of Green Building Costs— Block 225" 3D/I, 2003	Jim Ogden, associate
3	7 Group	John Boecker, Marcus Sheffer, principals
4	Albanese Organization	George Aridas, senior vice president
5	All Saints Parish	Tom Nutt-Powell, property committee chair
6	"Saving Resources," *Urban Land* (June 2001)	Anthony Bernheim
7	Arlington School District	Patty Kavanaugh
8	Arup Engineering	Cole Roberts, senior engineer; Megan Hemmerle, marketing coordinator
9	AtSite Real Estate	Chip Ranno, vice president
10	Ball State University	Kevin Kenyon
11	BNIM Architects	Brad Nies, associate director of elements; Jean Dodd, elements division
12	Boldt Construction, The Kubala-Washatko Architects	Theresa Lehman, LEED consultant; Joel Krueger, associate
13	Boora Architects	Heinz Rudolf, principal; Berthe Carroll, project assistant
14	Bordo International	Lisa Crowely, Cameron Brown
15	Browne Penland McGregor Stephens Architects	Randy Curry
16	Bruner Cott & Associates, Inc.	
17	Building Green High Performance Building Database	Michael Wentz and Nadav Malin
18	Burt Hill Kosar Rittleman Associates	
19	Calvin College	Frank Gorman, architect
20	Cherokee Investment Partners	Tom Darden, chief executive officer; Chris Wedding, sustainable planning and development associate
21	Christensen Corporation	Gary Christensen

Firm	Name
22 Cohos Evamy|Integratedesign	Naomi Minja, communications director
23 Colorado Department of Labor	Angie Fyfe, project manager
24 Confederation of Indian Industry	
25 Conseil Scolaire de District du Centre-Sud-Ouest, Enermodal Engineering Ltd.	Jordan Hoogendam
26 Cook + Fox Architects	Robert Fox, partner
27 Corporate Office Properties Trust	Thomas Fahs, development manager
28 Croxton Collaborative	Randy Croxton
29 DPR Inc.	Craig Greenough
30 DR&I Architects	Paul Brown
31 E4 Incorporated	Pam Lippe, environmental consultant to the Durst Organization
32 Elk River Area Schools	Ron Bratlie, director of business, operations, and construction
33 Enterprise Green Communities	Stockton Williams, senior vice president and chief strategy officer; Dana Bourland, Green Communities director; Jerone Gagliano; Kristen Karle
34 Environmental Protection Agency	Cathy Berlow
35 De Anza College	Pat Cornely, Julie Phillips
36 Fraunhofer-Institute for Building Physics	Dietrich Schmidt
37 Friends Committee on National Legislation	Maureen Brookes, communications program assistant
38 GAP mbH	Thomas Winkelbauer
39 Georgia State Parks & Historic Sites	David Freedman, Katie Berfefeld
40 Gerding Edlen Development	Jessy Olson, Renée Worme
41 GGLO Architects	Jonathon Hall, Michelle Rosenberger
42 Gossens Bachman Architects;	Jeff Stetter
43 Greater Toronto Airport Authority	Anthony Margiotta, Sheila McGuigan
44 Green Technologies FZCO, Dubai Airport Free Zone Authority	R. M. Harshini de Silva, sustainability analyst
45 Hamilton-Anderson	Paul Locher
46 Harford Community College	Katherine McGuire, grants manager
47 Harvard Green Campus Initiative	Leith Sharp, director; Andrea Ruedy, project coordinator
48 HMFH Architects	Doug Sacra, senior LEED architect
49 HOK	
50 Homeword	Heather McMillin
51 Housing Vermont and Burlington Community Land Trust	

	Firm	Name
52	inFORM Studio	Neil Chambers, director of sustainability
53	Integrated Architecture	
54	Ithaca College	
55	John O'Hara Associates	Randy Overton
56	Karlsberger Architecture Inc.	Tom Snearey
57	Kiss + Cathcart Architects	Jeff Miles, Colin Cathcart
58	Kulp Boecker Architects	
59	LD Astorino Co.	
60	Leeb Architects	Robert Leeb
61	Lend Lease	
62	LPA Group	
63	Mahlum Architects	Katrina Morgan, Bill Strong, Anne Schopf, principals
64	Margo Jones Architects	Margo Jones
65	Matsuzaki Architects, Inc.	Eva Matsuzaki
66	Melink	Jason Brown
67	Merrill Architects	Tim Merrill
68	Metropolitan Architects and Planners	
69	Missouri Department of Natural Resources	Dan Walker
70	Moseley Architects	Bryna Dunn, director of environmental planning and research; Gillian Rizy, environmental analyst
71	Mostue & Associates Architects	Iric Rex
72	National Association of Realtors	Joe Molinaro
73	Natural Resources Defense Council, Report on Robert Redford Building	Rob Watson, chief executive officer, Ecotech International; founding chair, U.S. Green Building Council
74	New York City Department of Design and Construction	Rebecca Massey and John Krieble
75	OWP/P	Kevin Hall and Rand Eckman, sustainable design
76	Paladino and Company	Brad Pease
77	PAPSA	Martha Sepulveda, operations manager
78	Perkins and Will Architects	Paula Vaughan, director of sustainability
79	Plymouth State University	Bill Crangle, vice president of financial affairs
80	PN Hoffman	Sean Seaman
81	Poudre School District	Stu Reeve, energy manager
82	PRP Architects	Dave Deppen

	Firm	Name
83	Renschler	Eric Truelove, company director of sustainable design services
84	Rocky Mountain Institute	James Scott Brew, principal
85	Ross Barney Architects	
86	Salem Engineering, MEP	Andy Shapiro, LEED and energy consultant
87	Seattle University	Michael George, associate vice president for facilities services
88	Serena Sturm Architects, Ltd.	
89	Sidwell Friends Middle School	Mike Saxenian, assistant head of school, chief financial officer
90	Sustainability Victoria, Commercial Office Building+B133 Energy Innovation Initiative	
91	Sustainable Design Consulting	Sandra Leibowitz Earley, principal; Beth Ridout, project consultant
92	Swinerton Incorporated	Grant French, corporate sustainability manager
93	TMP Architects	Eric Sassak
94	Tower Companies	Elizabeth Lisboa, Marnie Abramson
95	Twenhofel Middle School, Kenton County School District	Robert Lape, facilities director; Chris Baker, energy systems coordinator
96	Unitarian Universalist Church, Fresno	George Burman, project manager
97	University of British Columbia	Alison Aloisio, sustainable buildings advisor
98	Viikki Housing Development, City of Helsinki Economic and Planning Centre	Heikki Rinne, project manager
99	Department of Conservation, Wellington, New Zealand	
100	William T Moore Construction	Bill Moore, principal
101	Wolff, Lang, Christopher Architects, Inc., Resource Guide	
102	World Build	David Gottfried
103	Zimmerman Design Group	

APPENDIX C
GREEN BUILDING DATA SET

The Green Building Data Set appears on pages 200 to 209.

(See also www.islandpress.org/Kats)

Project name	City	State (if U.S.) or country	New (N) or renovation (R)	Source number[1]	Building type	LEED level (or equivalent)
Affordable multifamily housing						
Clara Vista Townhomes	Portland	OR	N	33	Affordable housing	Green Communities
Orchard Gardens	Missoula	MT	N	50	Affordable housing	Certified
Royal Building	Springfield	OR	N	33	Affordable housing	Green Communities
Southeast Phillips Creek	Milwaukee	OR	N	67	Affordable housing	Earth Advantage
Station Place Tower	Portland	OR	N	60	Affordable housing	Earth Advantage
Waterfront housing	Burlington	VT	N	51, 42, 86	Affordable housing	Certified
Correctional facilities						
Federal correctional institution 3	Butner	NC	N	70	Correctional	Certified
District chilled-water plant						
The Wafi City District Cooling Chilled Water Plant–DCCP ONE	Dubai	United Arab Emirates	N	44	District cooling plant	Gold
Health care facilities						
Boulder Community Foothills Hospital	Boulder	CO	N	104	Healthcare-Acute Care	Silver
Center for Discovery	Harris	NY	N	104	Healthcare-MOB/outpatient	Certified
Dell Children's Medical Center of Central Texas	Austin	TX	N	56	Health care	Platinum
Denver Health Pavilion for Women and Children	Denver	CO	N	104	Healthcare-Acute Care	Silver
Geisinger Gray's Woods Ambulatory Care Facility	State College	PA	N	104	Healthcare-MOB/outpatient	Gold
Geisinger Wyoming Valley, Critical Care Building	Wilkes-Barre	PA	N	104	Healthcare-Acute Care	Silver
Jersey Shore	Neptune	NJ	N	104	Healthcare-Acute Care	Silver
Lacks Cancer Center	Grand Rapids	MI	N	104	Healthcare-Acute Care	Certified
Metro Health	Wyoming	MI	N	104	Healthcare-Acute Care	Certified
Oregon Health Sciences University	Portland	OR	N	40	Health care	Platinum
Parrish Healthcare Center at Port St. John	Cocoa	FL	N	104	Healthcare-MOB/outpatient	Silver
Pearland Pediatrics	Pearland	TX	N	15	Health care	Certified
Providence Newberg Medical Center	Newberg	OR	N	104	Healthcare-Acute Care	Gold
St. Mary's Duluth Clinic	Duluth	MN	N	84	Health care	Gold
Spaulding Rehabilitation Hospital for Health Care	Boston	MA	N	78	Health care	Green Guide
Higher education						
Blackstone, Harvard	Cambridge	MA	R	16, 47	Higher education	Platinum
C. K. Choi Building	Vancouver	Canada	N	65	Higher education	Gold (equiv.)
Environmental Studies Center, De Anza College	Cupertino	CA	N	35, 8, 82	Higher education	Platinum
Ithaca College Gateway Building	Ithaca	NY	N	54	Higher education	Platinum
Joppa Hall	Bel Air	MD	R	46	Higher education	Silver
Landmark Center, Harvard University	Cambridge	MA	R	47	Higher education	Gold

Building square footage	Year constructed	Cost of building ($/sf)[2]	Green premium[3]	Percent reduction in energy use[4]	Modeled (M) or actual (A)	Energy use baseline[5]	Percent reduction in water use[6]	Percent construction and demolition waste diverted[7]	Percent recycled content	Percent local materials
53,900	2006	106	0.0%	73%	A	3	—	—	—	—
37,782	2005	170	0.0%	43%	M	—	—	—	—	—
38,105	[8]	153	3.0%	37%	M	—	43%	—	—	—
—	2005	—	3.5%	15%	M	4	—	80%	—	—
15,4359	2004	171	3.5%	—	M	—	—	—	—	—
40,000	2004	126	8.4%	40%	A	—	47%	—	7%	36%
53,0295	2005	186	0.3%	31%	M	—	34%	70%	6%	26%
—	2006	—	4.5%	30%	M	—	55%	—	—	—
154,000	2003	—	2.90%	28%	M	—	—	64%	19%	36%
28,000	2003	214.29	2.00%	24%	M	—	—	0.5	0.0535	0.2
470,000	2007	234	0.0%	50%	M	—	35%	97%	10%	50%
212,000	2006	297	—	19%	M	—	—	0.5246	0.4165	0.25
51,800	2008	404	0.50%	20%	M	—	21%	81%	15%	20%
121,465	2008	337	0.50%	21%	M	2	0.301	0.88	0.1	0.2
336,000	2009	320	0.00%	36%	M	2	30%	50% target	20% target	20%
170,000	2004	258	1.00%	0%	M	2	21%	98%	15%	21%
468,801	2007	260	0.00%	0%	M	—	0%	0.7075	0.0968	0.2337
402,400	2006	373	1.1%	40%	M	4	61%	97%	16%	23%
72,236	2006	209	3.80%	15%	M	—	30%	50%	50%	20%
10,387	2006	131	4.0%	13%	M	—	—	—	—	—
183,004	2006	385.25	0.50%	26%	M	4	20%	0.8091	0.25	0.3
240,000	2006	263	0.8%	18%	M	—	36%	77%	19%	42%
240,000	2007	479	2.5%	10%	M	2	—	50%	20%	20%
40,000		250	0.0%	45%	M	—	32%	12%	—	—
34,400	1996	150	0.0%	57%	M	—	—	50%	—	—
22,000	2006	330	5.5%	88%	M	—	44%	—	—	—
57,500	2008	284	1.8%	34%	M	2	89%	59%	20%	20%
77,357	2005	—	3.0%	24%	A	3	—	75%	25%	50%
42,000	2001	118	0.0%	—	M	—	20%	97%	10%	71%

Project name	City	State (if U.S.) or country	New (N) or renovation (R)	Source number[1]	Building type	LEED level (or equivalent)
Langdon Woods Residence Hall	Plymouth	NH	N	79	Higher education	Gold
Old Dominion University Engineering Building	Norfolk	VA	N	70	Higher education	Certified
Park Hall, Ball State University	Muncie	IN	N	10	Higher education	Silver
Rinker Hall at the University of Florida	Gainesville	FL	N	17	Higher education	Gold
Science II Replacement Building, California State University at Stanislaus	Turlock	CA	N	78	Higher education	Certified
Seattle University Student Center	Seattle	WA	N	87	Higher education	Certified
Seminar II, The Evergreen State College	Olympia	WA	N	63	Higher education	Gold
Stanford Energy and Environment Building	Palo Alto	CA	N	8	Higher education	Platinum
University of British Columbia Life Sciences Centre	Vancouver	Canada	N	97	Higher education	Gold
University of Scranton Campus Center	Scranton	PA	N	18	Higher education	Certified
Vincent and Helen Bunker Interpretive Center	Grand Rapids	MI	N	19	Higher education	Gold

K–12 schools

Project name	City	State	New (N)	Source	Building type	LEED level
Ash Creek Intermediate School	Monmouth	OR	N	13	K–12	Silver
Ashland High School	Ashland	MA	N	48	K–12	MA-CHPS
Baker Prairie Middle School	Canby	OR	N	13	K–12	Silver
Berkshire Hills	—	MA	N	48	K–12	MA-CHPS
Blackstone Valley Tech	—	MA	N	48	K–12	MA-CHPS
Clearview Elementary	Hanover	PA	N	3	K–12	Gold
Crocker Farm School	Amherst	MA	N	64	K–12	MA-CHPS
C-TEC	Newark	OH	N	3	K–12	Silver
Danvers	—	MA	N	48	K–12	MA-CHPS[9]
Dedham	—	MA	N	48	K–12	MA-CHPS
Detroit School of Arts	Detroit	MI	N	45	K–12	Certified
Fossil Ridge High School	Fort Collins	CO	N	81	K–12	Silver
Franklin Elementary	Kirkland	WA	N	63	K–12	AIA COTE award[10]
Hector Garcia Middle School	Dallas	TX	N	78	K–12	Certified
Hermitage Elementary	Virginia Beach	VA	N	70	K–12	Certified
Jeunes sans Frontières Secondary School	Brampton	Canada	N	25	K–12	Silver
Kersey Creek Elementary	Mechanicsville	VA	N	70	K–12	Silver
Melrose Middle School	Melrose	MA	N	48	K–12	MA-CHPS
Michael E. Capuano Early Childhood Education Center	Somerville	MA	N	48	K–12	Certified
Model Green School	Chicago	IL	N	75	K–12	Silver
Newton South High School	Newton	MA	N	30	K–12	Silver
North Clackamas High	Clackamas	OR	N	13	K–12	Silver
Prairie Crossing Charter School	Grayslake	IL	N	88	K–12	Silver
Punahou School	Honolulu	HI	N	55	K–12	Gold

Building square footage	Year constructed	Cost of building ($/sf)[2]	Green premium[3]	Percent reduction in energy use[4]	Modeled (M) or actual (A)	Energy use baseline[5]	Percent reduction in water use[6]	Percent construction and demolition waste diverted[7]	Percent recycled content	Percent local materials
95,000	2006	217	2.2%	58%	M	—	43%	80%	19%	40%
82,715	2005	138	0.0%	21%	A	—	25%	78%	25%	25%
174,111	2007	160	0.6%	39%	M	—	29%	87%	19%	27%
47,300	2003	137	2.8%	57%	M	—	—	75%	10%	20%
115,000	2007	389	5.0%	10%	M	4	20%	75%	10%	—
64,000	2005	—	1.0%	30%	M	—	—	—	—	—
168,000	2004	190	0.0%	25%	M	—	31%	79%	—	18%
160,000	2007	460	2.8%	36%	M	2	80%	—	—	—
561,521	2008	226	0.4%	29.6%	M	—	9.65%	78%	12%	16%
118,000	2007	271	0.6%	23%	M	—	10%	—	—	—
4,500	—	289	8.1%	—	M	—	—	—	—	—
58,000	2002	124	0.0%	30%	M	4	20%	—	—	—
202,465	2005	195	1.9%	29%	M	—	—	—	—	—
135,000	2006	171	3.1%	39%	M	4	—	—	65%	—
78,000	2004	172	4.0%	34%	M	—	0%	—	—	—
277,263	2005	130	0.9%	32%	M	—	12%	—	—	—
43,450	2002	155	1.3%	59%	M	—	39%	90%	—	—
69,339	2001	—	1.1%	32%	M	—	62%	—	—	—
329,140	—	112	0.5%	23%	M	—	45%	95%	—	—
148,000	2005	165	3.8%	23%	M	—	7%	—	—	—
130,100	2006	202	2.9%	29%	M	—	78%	—	—	—
286,000	2005	205	0.1%	23%	M	—	2%	57%	—	—
—	—	—	0.0%	—	M	—	—	—	—	—
56,000	2006	176	3.0%	35%	M	—	—	—	—	—
172,169	2007	140	0.6%	20%	M	—	59%	—	10%	20%
85,500	2004	106	0.0%	10%	A	3	33%	68%	14%	24%
8,463	2007	145	4.5%	47%	M	5	31%	91%	24%	48%
76,316	2006	164	0.0%	31%	M	—	41%	77%	32%	62%
—	2006	—	1.4%	20%	M	—	20%	—	—	—
80,000	2003	165	3.6%	41%	M	—	0%	—	—	—
120,000	—	157	2.0%	29%	M	—	35%	—	—	—
399,000	—	89	1.0%	30%	M	—	20%	—	—	—
268,269	2002	113	0.3%	39%	M	2	20%	50%	—	—
13,613	2005	195	3.4%	43%	M	2	43%	—	—	—
223,286	2004	250	6.3%	43%	M	—	50%	75%	—	—

Project name	City	State (if U.S.) or country	New (N) or renovation (R)	Source number[1]	Building type	LEED level (or equivalent)
Sidwell Friends Middle School	Washington	D.C.	N	89	K–12	Platinum
Skyline High School	Ann Arbor	MI	N	93	K–12	Certified (equiv.)
Summerfield Elementary	Neptune	NJ	N	3	K–12	Gold
The Dalles Middle School	The Dalles	OR	N	13	K–12	Gold
Third Creek Elementary	Iredell	NC	N	70	K–12	Gold
Twenhofel Middle School	Fort Wright	KY	N	95	K–12	Silver
Twin Lakes Elementary	Elk River	MN	N	32	K–12	Gold
Twin Valley Elementary	Elverson	PA	N	3	K–12	Silver
Washington-Lee High School	Arlington	VA	N	91, 7	K–12	Silver
Washington Middle School	Olympia	WA	R	63	K–12	Gold
Whitman Hanson	—	MA	N	48	K–12	MA-CHPS
Williamstown Elementary School	Williamstown	MA	N	64	K–12	MA-CHPS
Woburn High School	Woburn	MA	N	48	K–12	MA-CHPS
Woodward Academy classroom	College Park	GA	N	78	K–12	Silver
Woodward Academy dining	College Park	GA	N	78	K–12	Certified
Wrightsville Elementary	Wrightsville	PA	N	3	K–12	Silver
Laboratory						
Federal lab	—	GA	N	78	Lab	Gold
Senator William X. Wall Experiment Station	Lawrence	MA	R	78	Lab	Gold
Three-story laboratory	Orlando	FL	N	78	Lab	Silver
Office						
4 Times Square	New York	NY	N	31	Office	Silver (equiv.)
30 The Bond	Sydney	Australia	N	61	Office	Green Star
40 Albert Road	Melbourne	Australia	R	90	Office	—
Aldo Leopold Center	Baraboo	WI	N	12	Office	Platinum
American Speech-Language-Hearing Association national office	Rockville	MD	N	91, 9	Office	Gold
Banner Bank Building Benefit Fund	Boise	ID	N	21	Office	Platinum
Bordo International	Victoria	Australia	N	14	Office	Green Star
Brengel Tech Center	Milwaukee	WI	N	103	Office	Gold
California Environmental Protection Agency headquarters	Sacramento	CA	N		Office	
Center for Sustainable Building—ZUB	Kassel	Germany	N	36	Office	—
Cherokee main office	Raleigh	NC	R	20	Office fit-out	Platinum
CII-Sohrabji Godrej Green Business Centre	Hyderabad	India	N	24	Office	Platinum
Conservation House	Wellington	New Zealand	R	99	Office	Green Star
Development Resource Center	Chattanooga	TN	N	28	Office	Gold
DPR/ABD Office Building	Sacramento	CA	N	102, 29	Office	Silver
East End Complex	Sacramento	CA	N	2	Office	Gold

Building square footage	Year constructed	Cost of building ($/sf)[2]	Green premium[3]	Percent reduction in energy use[4]	Modeled (M) or actual (A)	Energy use baseline[5]	Percent reduction in water use[6]	Percent construction and demolition waste diverted[7]	Percent recycled content	Percent local materials
72,200	2006	388	9.6%	60%	M	—	93%	60%	11%	78%
380,564	2007	—	1.0%	35%	M	—	20%	—	—	—
99,723	—	221	0.8%	32%	M	—	35%	—	—	—
96,500	2002	130	0.5%	46%	M	—	20%	—	—	—
89,695	2002	106	1.5%	34%	A	1	32%	58%	66%	54%
112,000	2006	205	10.1%	29%	A	3	30%	—	—	—
98,000	2007	190	0.0%	52%	M	—	—	—	—	—
70,160	2004	163	1.5%	49%	M	—	42%	77%	—	—
350,000	2009	273	2.4%	26%	M	—	46%	50%	27%	32%
100,000	2006	170	3.0%	25%	M	—	40%	90%	—	—
234,500	2005	179	1.5%	35%	M	—	38%	—	—	—
80,000	2002	—	0.0%	31%	M	—	0%	—	—	—
340,000	2006	196	3.1%	30%	M	—	50%	—	—	—
—	2002	—	0.0%	31%	M	—	23%	—	—	—
—	2003	—	0.1%	23%	M	—	25%	—	—	—
80,400	2003	120	0.4%	30%	M	—	23%	85%	—	—
136,350	2006	210	0.6%	20%	M	—	—	57%	22%	46%
33,600	2007	357	5.0%	21%	M	2	30%	75%	20%	20%
175,000	2008	364	1.8%	16%	M	—	20%	75%	5%	10%
—	1999	—	7.5%	—	M	—	—	—	—	—
—	—	—	—	—	M	5	—	—	—	—
—	2005	—	—	—	M	5	—	—	—	—
8,844	2007	293	12.5%	115%	M	—	55%	75%	10%	20%
137,500	2007	230	6.5%	17%	M	—	42%	87%	13%	45%
180,000	2006	122	0.0%	51%	A	—	80%	—	20%	20%
31,056	2004	66	0.0%	68%	M	3	0%	—	—	—
460,000	2000	125	0.0%	35%	A	—	20%	—	—	—
950,000	2001	—	1.6%	34%	A	—	20%	—	—	—
13,992	2002	230	—	77%	A	3	—	—	—	—
22,510	2007	68	10.2%	26%	M	2	45%	86%	40%	12%
20,000	2003	—	—	55%	M	—	>30%	—	—	—
—	2007	—	—	40%	M	5	60%	50%	—	—
—	2001	—	1.0%	0%	M	—	0%	—	—	—
52,300	2003	0	0.9%	45%	M	—	75%	—	—	—
—	2003	—	6.4%	45%	M	—	30%	75%	50%	20%

Project name	City	State (if U.S.) or country	New (N) or renovation (R)	Source number[1]	Building type	LEED level (or equivalent)
Energy Resource Center	Downey	CA	N	101	Office	Certified
Environmental Protection Agency Region 7 Headquarters	Kansas City	KS	N	1	Office	Silver
Friends Committee on National Legislation Building	Washington	D.C.	R	37	Office	Silver
Heifer International Center	Little Rock	AR	N	11	Office	Platinum
Lewis and Clark State Office Building	Jefferson City	MO	N	11, 69	Office	Platinum
Melink Headquarters	Cincinnati	OH	N	66	Office	Gold
National Association of Realtors Headquarters	Washington	D.C.	N	72	Office	Silver
National Business Park 306	Annapolis Junction	MD	N	27, 91	Office	Silver
National Business Park 318	Annapolis Junction	MD	N	27, 91	Office	Gold
Natural Resources Defense Council, Santa Monica Office—Robert Redford Building	Santa Monica	CA	R	73	Office	Platinum
Nidus Center of Science	Creve Coeur	MO	N	49	Office	Silver
Office of Emergency Management	Brooklyn	NY	N	74	Office	Silver
One Bryant Park (Bank of America Building)	New York	NY	N	26	Office	Platinum
PA DEP California	California	PA	N	3	Office	Gold
PA DEP Cambria	Ebensburg	PA	N	3	Office	Gold
PA DEP Southeast	Norristown	PA	N	3	Office	Gold
PAPSA Monterrey	—	Mexico	R	77	Office	Gold
Paul Wunderlich-Haus	Eberswalde	Germany	R	38	Office	—
PCL Centennial Learning Centre	Edmonton	Canada	N	22	Office	Gold
Pennsylvania Department of Environmental Protection (PA DEP) South Central Regional	Harrisburg	PA	N	58	Office	Certified
Pier One	San Francisco	CA	N	6	Office	Certified
PNC Firstside Center	Pittsburgh	PA	N	59	Office	Silver
Potomac Yard One and Two	Arlington	VA	N	91, 68, 34	Office	Gold
Queens Botanical Garden	Flushing	NY	N	74	Office	Platinum
RenewAire renovation	Madison	WI	R	83	Office	Silver
Swinerton headquarters building	San Francisco	CA	R	92	Office	Gold
Tower Building	Rockville	MD	N	94	Office	Certified (equiv.)
Toyota Motor Sales, South Campus Headquarters	Torrance	CA	R	62	Office	Gold
Villa Park Police Station	Villa Park	IL	N	88	Office	Silver
Wisconsin Electrical Employees	Madison	WI	N	83	Office	Certified
Performance space						
The Armory (Block 3)	Portland	OR	R	40	Performance space	Platinum
Other public buildings						
Brooklyn Children's Museum	Crown Heights	NY	R	74	Public	Silver

Building square footage	Year constructed	Cost of building ($/sf)[2]	Green premium[3]	Percent reduction in energy use[4]	Modeled (M) or actual (A)	Energy use baseline[5]	Percent reduction in water use[6]	Percent construction and demolition waste diverted[7]	Percent recycled content	Percent local materials
—	1995	—	0.0%	45%	M	1	—	—	—	—
217,000	—	147	0.0%	0%	M	—	—	—	—	—
9,775	2005	379	1.9%	48%	M	—	50%	—	—	—
89,992	2005	188	13.8%	55%	M	—	50%	96%	22%	72%
120,000	2005	147	4.0%	60%	M	—	—	63%	50%	50%
30,000	2005	87	9.6%	50%	M	—	50%	—	—	—
105,000	2004	295	2.0%	39%	M	—	52%	—	—	—
162,000	2006	24, 130, 863	1.1%	—	M	—	—	—	—	—
125,000	2004	168	2.9%	37%	M	—	39%	67%	—	—
15,000	2003	340	13.6%	64%	M	—	60%	90%	—	—
41,233	1999	247	3.4%	38%	M	1	20%	—	—	—
58,000	2006	759	1.4%	4%	M	—	33%	62%	17%	23%
2,000,000	2008	650	2.0%	50%	M	—	50%	83%	35%	40%
—	2003	—	1.7%	40%	M	—	41%	—	—	—
36,000	2000	93	1.2%	66%	M	—	33%	—	—	30%
—	2003	—	0.1%	41%	M	—	20%	—	—	—
4,500	2007	130	12.0%	29%	M	2	20%	—	—	—
52,487	2007	172	0.0%	60%	M	3	0%	—	—	—
—	2006	—	—	38%	M	5	43%	96%	20%	—
—	1998	—	1.0%	20%	M	1	33%	—	—	—
52,300	2001	—	0.7%	10%	M	4	—	—	—	—
—	2000	160	0.3%	33%	M	—	—	—	—	—
654,000	2006	—	4.0%	21%	M	—	42%	71%	27%	61%
13,900	2007	1,007	7.7%	21%	M	—	64%	90%	19%	54%
37,096	2005	—	1.5%	40%	M	—	—	82%	21%	78%
67,000		91	1.8%	12%	M	4	40%	88%	28%	—
276,000	2001	188	1.3%	13%	M	3	—	—	—	—
624,000	2003	63	0.0%	59%	M	4	94%	95%	0%	0%
16,000	2004	216	4.8%	33%	M	2	26%	79%	16%	85%
12,000	2005	—	3.0%	35%	M	—	—	57%	20%	26%
51,000	2006	500	0.3%	30%	M	—	89%	95%	25%	45%
90,000	2007	433	0.5%	44%	M	4	>30%	75%	25%	20%

Project name	City	State (if U.S.) or country	New (N) or renovation (R)	Source number[1]	Building type	LEED level (or equivalent)
Chesterfield Community Development Customer Service Center	Chesterfield	VA	N	70	Public	Silver
Colorado Department of Labor and Employment	Denver	CO	N	23	Public	Certified
Fire and Emergency Services Training Institute, Greater Toronto Airport Authority	Mississauga	Canada	N	43	Public	Silver
Remsen Yard	Brooklyn	NY	R	74	Public	Silver
Sunrise Yard	Ozone Park	NY	N	74	Public	Gold
Sweetwater Creek State Park Visitor Center	Lithia Springs	GA	N	39	Public	Platinum
Weeksville New Education Building	Brooklyn	NY	N	74	Public	Gold
Religious assembly						
All Saints Parish	Brookline	MA	R	5	Religious assembly	Gold
Jewish Reconstructionist Congregation synagogue	Evanston	IL	N	85	Religious assembly	Gold
Keystone Community Church	Ada	MI	N	53	Religious assembly	Certified
Unitarian Universalist Church of Fresno	Fresno	CA	N	96	Religious assembly	Silver
Residential						
4115 Avenue H	Austin	TX	R	100	Residential	Austin 5-Star
Blair Towns	Silver Spring	MD	N	94	Residential	Certified
Eco-Vikki	Helsinki	Finland	N	98	Residential	—
Linden Street Project	Somerville	MA	N	71	Residential	Certified (equiv.)
Takoma Village Cohousing	Washington	D.C.	N	91, 80	Residential	Certified (equiv.)
The Alta	Washington	D.C.	N	91	Residential	Certified
The Solaire	New York	NY	N	4	Residential	Gold
Multifamily/ mixed-use						
Alcyone	Seattle	WA	N	41	Residential/mixed-use	Certified
Allegiant	North Myrtle Beach	SC	N	52	Residential/mixed-use	Gold
Block 1 (Whole Foods)	Portland	OR	N	40	Residential/mixed-use	Silver
Brewery Block 2	Portland	OR	N	40	Residential/mixed-use	Gold
Broadway Crossing	Seattle	WA	N	41	Residential/mixed-use	Silver
M Financial Plaza (Brewery Block 4)	Portland	OR	N	40	Residential/mixed-use	Gold
Stone Way Apartments	Seattle	WA	N	41	Residential/mixed-use	Silver
The Henry (Block 3)	Portland	OR	N	40	Residential/mixed-use	Gold
The Lee Residence	New York	NY	N	57	Residential	Silver
The Louisa (Block 5)	Portland	OR	N	40	Residential/mixed-use	Gold
Retail						
Bronx Zoo Lion House	Bronx	NY	R	74	Zoo exhibit	Gold
PNC Green Branch Adams Township	Mars	PA	N	76	Retail (branch bank)	Gold
PNC Green Branch Ashburn Crossroads	Ashburn	VA	N	76	Retail (branch bank)	Silver
PNC Green Branch Fairfield Township	Fairfield Township	OH	N	76	Retail (branch bank)	Gold

Building square footage	Year constructed	Cost of building ($/sf)[2]	Green premium[3]	Percent reduction in energy use[4]	Modeled (M) or actual (A)	Energy use baseline[5]	Percent reduction in water use[6]	Percent construction and demolition waste diverted[7]	Percent recycled content	Percent local materials
87,010	2006	129	0.0%	0%	M	—	36%	—	24%	60%
40,000	2004	100	0.0%	21%	M	—	26%	52%	27%	34%
30,185	2007	460	1.5%	31%	M	5	20%	83%	19.625%	31.5%
33,000	2008	301	2.30%	19%	M0	2	40%	>75%	10%	10%
21,500	2008	758	2.9%	66%	M	—	33%	75%	10%	20%
—	2006	—	4.1%	34%	M	—	80%	—	—	—
19,500	2009	718	0.0%	23%	M	—	40%	75%	5%	21%
26,400		7	8.4%	25%	M	—	39%	—	—	—
32,000	2008	228	3.4%	30%	M	—	—	—	—	—
31,000	2004	116	3.0%	35%	M	—	—	—	—	—
11,000	2007	336	4.0%	34%	M	4	43%	100%	30%	
2,400	2007	150	10.0%	22%	M	4	50%	88%	50%	5%
87,613	2003	194	3.0%	35%	M	—	20%	—	40%	63%
—	2004	—	5.0%	20%	A	5	22%	—	—	—
41,000	2002	159	0.5%	43%	M	—	25%	—	—	—
51,000	2001	119	1.4%	—	M	—	0%	—	—	—
—	2006	—	1.5%	15%	M	—	—	0%	10%	40%
357,000	2003	321	18.0%	35%	M	—	50%	93%	60%	67%
149,697	2004	71	0.0%	33%	A	—	15%	98%	7%	27%
1,300,000	2009	250	2.0%	45%	M	2	55%	75%	20%	20%
158,000	2002	234	0.8%	24%	M	—	—	95%	—	40%
248,000	2002	233	0.5%	22%	M	—	8%	95%	50%	51%
52,600	2007	120	0.1%	19%	M	2	41%	84%	14%	21%
295,000	2003	234	0.8%	22%	M	—	25%	96%	—	—
75,129	2007	105	0.5%	40%	M	—	33%	90%	10%	20%
220,000	2004	185	0.4%	57%	A	—	5%	95%	50%	20%
99,000	2009	291	4.9%	28%	M	4	42%	50%	20%	10%
290,000	2005	123	1.0%	44%	M	—	35%	95%	20%	27%
29,011	2006	1,069	4.4%	56%	M	—	53%	96%	10%	20%
3,650	2006	—	0%	50%	M	—	58%	50%	10%	50%
3,650	2006	—	0%	40%	M	—	43%	75%	10%	50%
3,650	2006	—	0%	45%	M	—	58%	50%	10%	50%

Project name	City	State (if U.S.) or country	New (N) or renovation (R)	Source number[1]	Building type	LEED level (or equivalent)
PNC Green Branch Greengate	Greenburg	PA	N	76	Retail (branch bank)	Silver
PNC Green Branch Hamilton	Hamilton	NJ	N	76	Retail (branch bank)	Gold
PNC Green Branch Lower Macungie	Macungie	PA	N	76	Retail (branch bank)	Silver
PNC Green Branch Newtown	Newtown	PA	N	76	Retail (branch bank)	Silver
PNC Green Branch Rockaway Town Square	Dover	NJ	N	76	Retail (branch bank)	Gold
PNC Green Branch Tyler Center	Louisville	KY	N	76	Retail (branch bank)	Silver
PNC Green Branch Valley Square	Warrington	PA	N	76	Retail (branch bank)	Gold

Notes: All fields in the original data-collection sheet are not displayed in the table; those survey questions that did not result in enough responses to warrant inclusion in the analysis are omitted.

1 Source numbers correspond to the data-source list in appendix B.

2 Includes design and construction costs; excludes land costs.

3 Defined as "cost difference between actual building and the same building constructed using conventional building practices/without green features." For buildings that received incentives or grants for green building, the green premium is after incentives.

4 The percent reduction in energy use was determined through a modeled comparison with a conventional building or through a comparison with the actual energy use in the completed building.

Building square footage	Year constructed	Cost of building ($/sf)[2]	Green premium[3]	Percent reduction in energy use[4]	Modeled (M) or actual (A)	Energy use baseline[5]	Percent reduction in water use[6]	Percent construction and demolition waste diverted[7]	Percent recycled content	Percent local materials
3,650	2006	—	0%	45%	M	—	43%	75%	10%	50%
3,650	2006	—	0%	50%	M	—	58%	75%	10%	50%
3,650	2006	—	0%	45%	M	—	43%	50%	10%	50%
3,650	2006	—	0%	45%	M	—	43%	75%	10%	50%
3,650	2005	—	0%	45%	M	—	58%	75%	10%	50%
3,650	2006	—	0%	40%	M	—	43%	75%	10%	50%
3,650	2005	—	0%	45%	M	—	58%	75%	10%	50%

5 Except where noted, all reported energy-use reductions are based on the ASHRAE 90.1 1999 baseline. Key to numbers: 1 = based on an earlier version of ASHRAE 90.1; 2 = based on ASHRAE 90.1 2001 or 2004; 3 = based on energy use in conventional nearby buildings of the same type, as determined by the data source; 4 = based on a state energy code (e.g., California Title 24, Oregon Energy Code); 5 = based on a non-U.S. energy standard.

6 Water-use reductions based on modeling of green building compared with Energy Policy Act of 1992 requirements for all U.S. buildings.

7 Figures based on the method of calculation used in LEED materials credits (see www.usgbc.org).

8 Cells marked with a dash (—) indicate that the information was not available.

9 MA-CHPS is the Massachusetts version of the California High-Performance School Standard.

10 AIA COTE is the American Institute of Architects Committee on the Environment.

APPENDIX D
COMPARISON OF DATA SET TO
LEED—NEW CONSTRUCTION BUILDINGS

Cost and benefit estimates presented in this book are based on the reported cost premiums and energy- and water-use reductions among the approximately 170 recent green buildings shown in appendix C. Comparing the performance and general characteristics of these buildings to the U.S. Green Building Council's records of LEED-certified buildings indicates that the study data set represents a slightly higher-performing population than LEED-certified buildings generally. As shown in figure D.1, buildings in the data set generally had higher reported energy savings than buildings certified under the LEED for New Construction standards. Figure D.2 shows that the proportions of Gold and Plat-

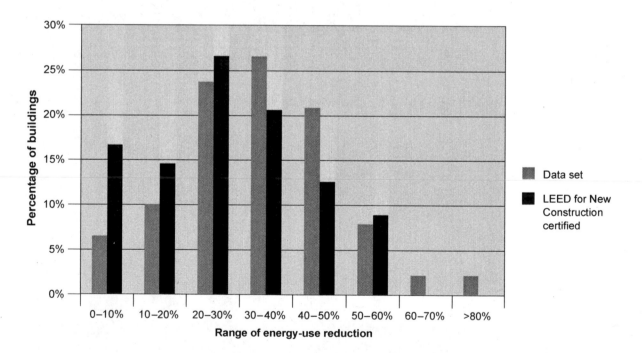

FIGURE D.1 Energy-Use Reductions in LEED for New Construction Buildings versus Buildings in the Study Data Set

Notes: (1) Range of energy savings for LEED for New Construction (LEED-NC) buildings was inferred based on the number of points earned in Energy and Atmosphere Credit 1. (2) LEED-NC 2.1 does not award extra points for energy-use reductions greater than 60%. (3) Some of the LEED-NC buildings in the 50% to 60% energy-use-reduction category may achieve savings greater than 60%.

inum buildings were higher in the data set than among LEED-certified buildings generally. The data set also has a higher proportion of schools and a smaller proportion of mixed-use projects than LEED-certified buildings generally.[1]

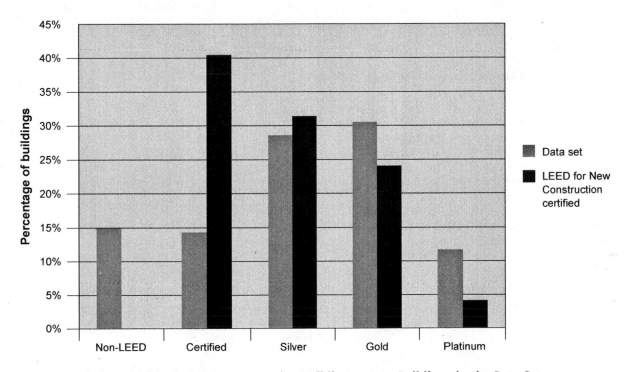

FIGURE D.2 LEED Level in LEED–New Construction Buildings versus Buildings in the Data Set
Note: Most non-LEED projects in the data set are roughly equivalent to LEED-certified or LEED Silver buildings.

APPENDIX E
BASELINES USED IN COST AND BENEFITS ESTIMATES

	Method of estimation	Baseline for nongreen building	Method of estimating $/sf impact in green buildings
Cost-benefit area	Architect (or other data source) reporting green cost estimate based on knowledge of specific project requirements	Same building without green features	Data set: Reported green premium (%) multiplied by total building cost
Energy savings	Comparison of projected or actual energy use in green building to a model of the same building that meets minimum code requirements (e.g., ASHRAE 90.1 2004)	Building with average U.S. expenditures for building type, based on Commercial Building Energy Consumption Survey (CBECS)[1]	Data set: Reported percentage reduction in energy use (from ASHRAE 90.1 standard), multiplied by average energy expenditures by building type (from CBECS), adjusted for inflation
Indirect energy savings	Assumption of natural gas and electrcity market price elasticity from reduced energy use	CBECS average energy expenditures	Assumed price impact of 25% of direct energy savings, based on referenced studies of demand price elasticity
Emissions reduction	Based on emissions intensity of fossil-fuel energy generation	CBECS average energy usage	Data set: Reported percentage reduction in energy use, multiplied by average emissions intensity by energy source (assumed 65% electric, 35% natural gas mix) for the U.S.
Water savings	Comparison of projected or actual energy use to a model of the same building that meets the code baseline	Average water use by building type: CIEUW survey;[2] average water rates: Raftelis Inc. national survey[3]	Data set: Reported percent reduction in water from fixture flow requirements in Energy Policy Act of 1992
Health improvements	Illustrative estimates as described in Part I.	Average health expenditures from poor indoor air quality	See Fisk[4] and Center for Building Performance and Diagnostics for additional information[5]
Employment	Input-output model of total economic impacts of green building	Model run using baseline inputs for building cost and energy and water expenditures	Model run using average green building cost and energy and water expenditures in data set

Notes:
1 Energy Information Administration (EIA), "Commercial Buildings Energy Consumption Survey," 2003 (www.eia.doe.gov/emeu/cbecs/contents.html).
2 Benedykt Dziegielewski et al., "Commercial and Institutional End Uses of Water," American Water Works Association, 2000.
3 Raftelis Financial Consultants, "National Water Rate Survey," 2006. Provided by Peiffer Brandt.
4 W. J. Fisk, "Health and Productivity Gains from Better Indoor Environments and Their Implications for the U.S. Department of Energy," 1999 (available at www.rand.org/scitech/stpi/Evision/Supplement/fisk.pdf).
5 See http://cbpd.arc.cmu.edu/bids.

APPENDIX F
ISSUES IN RESEARCHING THE COST OF GREEN BUILDING

Although the cost associated with green building is a topic of major interest among building-industry professionals, and among decision makers at all levels, the question "How much does it cost to build green?" poses inherent challenges to researchers. Buildings are complex, unique systems, consisting of multiple components that are affected by the site and by occupant activities and requirements.

Site characteristics—including topography, hydrology, ecology, history, and access to public transportation and services—may render entire LEED points impossible to achieve, or may make them achievable for any building, regardless of green objectives. Differences in local building codes, and in the availability of green materials, recycling facilities, and contractors who are familiar with green design, can change the feasibility and cost of green design. The relative costs of a green versus nongreen retrofit or renovation are likely to be different from the relative costs of a new green versus nongreen building. There are 20 renovations in the data set, ranging from office fit-outs to major renovations and additions; as discussed in section 1.2, "The Cost of Building Green," renovations typically experience slightly higher green cost premiums. Building systems considered innovative a few years ago may soon be considered standard practice, and may already be standard for certain owners or institutions. Green standards, such as LEED, also evolve—so the threshold of performance required to achieve green certification changes over time, further complicating research on the cost of green.

Many of our data sources reflected on the inherent difficulty in identifying costs associated with green design, and a few objected to the very notion of isolating a green premium. For example, Bryna Dunn, of Moseley Architects, said, "We haven't been able to assign a 'green premium' to many of our projects, because there really doesn't seem to be one. Or if there is, it is usually addressed within the already established project budget so nobody takes the time to break it out." Gary Christensen, the owner of the Banner Bank Building, commented that

> "Green Premium" really is the wrong concept—sorry. The real value to a developer/property owner is in the difference between market value and cost—the profit margin. My profit margin [on the green Banner Bank Building] is higher for sure. You really cannot separate the two discussions. People want to isolate the discussion to costs only, and they raise their hackles when they see higher cost totals. You really have to quantify how the energy savings and other savings translate into higher value, both now and in the future.

We asked our data sources—largely architects with intimate knowledge of project components, design process, and costs—to judge whether a particular material or building system should be considered green and what appropriate baseline costs for comparison would be, as well as to identify any tradeoffs or savings that may have also been involved. We asked that all additional costs or cost savings associated with achieving green design goals, including LEED certification costs, be included. While most green building projects do not conduct detailed cost analyses comparing their design to conventional processes, almost all face concerns about cost from owners, and must at some point consider and answer the question of how much it cost to green the project. For this reason, almost all architects report careful attention to keeping buildings and systems cost-effective through integrated design.

Conversations with architects indicate that prior experience with green building can significantly reduce the cost of building green. The number of individuals with such experience is growing rapidly, as is evidenced by the rapid increase in LEED Accredited Professionals (a professional credential indicating competence and familiarity with the LEED standard) to over 100,000.[1] This growing industry knowledge, along with the declining price of green products, materials, and systems, might be expected to lead to a decrease in reported green premiums over time. At the same time, however, the rapid expansion of green building means that a high percentage of those currently involved with green building projects are new to green design. Further, targets for reducing energy use for climate-change mitigation and energy independence can be expected to help to drive a toughening of green standards over the coming decades. For instance, in the spring of 2007, the U.S. Green Building Council approved a new requirement that all certified projects earn a minimum of two points in the Energy and Atmosphere category—effectively meaning that all future green buildings will achieve a minimum of 14% energy savings when compared with conventional design. Thus, it is not surprising that it is difficult to discern a change in green premiums experienced over the past five to ten years. For instance, the approximately 2% green premium reported by data sources in this study is in line with the findings of a 2003 study, "Costs and Financial Benefits of Green Buildings," despite the fact that much growth in green building has occurred in the intervening years. Green cost premiums may in fact remain constant over time, as experience with green design and the cost-effectiveness of efficient systems grows along with increasing thresholds for building performance.

APPENDIX G
COST OF ENERGY-EFFICIENCY AND RENEWABLE-ENERGY MEASURES

Table G.1 shows the reported net additional spending on energy efficiency, renewables, and green power compared with conventional buildings for 12 buildings from the data set for which detailed cost figures on energy measures were available.

TABLE G.1 ADDITIONAL INVESTMENT IN EFFICIENCY AND RENEWABLES IN GREEN BUILDINGS

Building	Total cost of energy-efficiency measures	Cost of on-site renewables and green power	Building size (s/f)	Net additional investment in energy efficiency and renewable energy compared with nongreen building ($/sf)
Brewery Block 2[1]	$258,284	Not applicable (N.a.)	248,000	$1.04
The Henry[1]	$299,533	N.a.	220,000	$1.36
M Financial Plaza (Brewery Block 4)[1]	$336,847	$309,066	295,000	$2.19
The Louisa (Brewery Block 5)	$377,009	N.a.	290,000	$1.30
Oregon Health Sciences University Center for Health and Healing[1]	$975,299	$886,000	402,400	$4.63
National Business Park 306	$57,000	N.a.	162,000	$0.35
National Business Park 318	$33,100	N.a.	125,000	$0.26
Washington-Lee High School	$587,500	$35,000	350,000	$1.78
Queens Botanical Garden	$233,070	$151,912	13,900	$27.70
Brooklyn Children's Museum[1]	$435,562	$150,000	90,000	$6.51
Office of Emergency Management[1]	$150,000	N.a.	58,000	$2.59
Stanford Energy and Environment Building[1]	$1,070,000	$100,000	160,000	$7.31

1 Reported costs for these buildings do not include the slight additional cost of commissioning, which is typically under $3/sf of building space.

APPENDIX H
ENERGY-USE BASELINES AND STANDARDS

Estimating energy savings in green buildings requires choosing a baseline level of performance against which green building energy use can be compared. As a baseline to calculate energy savings for LEED energy-performance credits, most LEED buildings use the American Society of Heating, Refrigerating, and Air-Conditioning Engineers' (ASHRAE) standard 90.1, which sets out minimum requirements for the efficiency of building systems. During the design process, engineers develop computer models of the building design and systems, which allow the building team to compare the energy impacts of alternative designs and technologies. These models also make it possible to compare projected energy use in a green building with a model of the same building that meets the minimum efficiency requirements of ASHRAE 90.1.

For the purposes of our study, buildings meeting the minimum requirements in ASHRAE 90.1 were considered to be typical of conventional, nongreen buildings; modeled reductions from ASHRAE 90.1 are therefore assumed to be the savings achieved by green buildings in comparison to conventional design. This is a simplifying assumption for several reasons. First, actual energy use in buildings can vary from projected energy use, although over a large number of buildings, models have been shown to provide a good prediction of actual energy use. (Roughly 10% of the buildings in the data set reported energy savings based on actual, post-occupancy records of energy use; the remainder reported savings based on energy modeling.) Second, ASHRAE 90.1 is an evolving standard; most buildings in the data set use the 1999 version of the standard, but some use the more stringent 2001 and 2004 versions (see appendix C). We did not adjust the reported energy-use reductions in the data set to account for the differences in the standards.

LEED and other green standards continue to evolve to reflect improvements in science, technology, and design. Moreover, green standards are based on performance standards that are also evolving—and generally becoming more rigorous (e.g., requiring lower adverse health and environmental impacts). ASHRAE 90.1, which covers commercial building efficiency and is revised every three years, offers an important example of such shifts. The 2010 version of ASHRAE 90.1 will require substantially greater efficiency than the 2007 version.

The calculations that would be needed to normalize performance data against different baselines would also suggest a level of precision that is not possible, given the diversity of data sources and building types and locations. Lastly, ASHRAE 90.1 is not a benchmark based on a large-scale survey of energy use in existing buildings, but a set of guidelines for the efficiency of building systems and a methodology for calculating appropriate energy use in buildings. Recent studies indicate that buildings designed to meet ASHRAE 90.1 1999 requirements may provide a good approximation of energy use in conventional buildings; that is, the energy use in buildings built to minimum ASHRAE 90.1 1999 standards is roughly equivalent to the average energy use of existing buildings (see appendix I). The choice of energy-use baselines for future cost-benefit analyses should be based on the evolving efficiency of existing building stock, current energy codes, and the goals of the particular analysis or benchmarking effort.

Despite the imprecision inherent in energy-savings estimates, ASHRAE 90.1 remains a broadly useful and reasonable baseline for this cost-benefit analysis. Recent studies have indicated that buildings designed to ASHRAE 90.1 are as efficient as, or more efficient than, typical existing buildings.[1] The choice of ASHRAE is also a practical one, since data on energy use compared with ASHRAE 90.1 are available for a large number of green buildings because of the LEED credits requiring energy modeling based on the standard. Other more detailed studies of energy use can add depth to the data collected here by analyzing utility records of energy use in green buildings compared to ASHRAE 90.1 or benchmarking tools, such as Energy Star (see appendix I).

APPENDIX I
VERIFYING THE ENERGY PERFORMANCE
OF LEED BUILDINGS

Reducing energy use is a major goal of green building design and construction.[1] All LEED buildings undergo full commissioning of energy systems, and most go through energy modeling. While these steps significantly reduce post-occupancy problems with mechanical systems and facilitate energy-efficient design, every building inevitably faces changes in operations and occupancy patterns over its lifetime, and many buildings experience unexpected technical issues, raising a number of important questions:

- Are green buildings realizing their energy-performance goals?
- Do energy models provide a reasonable means of assessing future energy use in green buildings?
- How does actual energy use in green buildings compare with energy use in existing conventional buildings?

In 2007, the New Buildings Institute (NBI) and the U.S. Green Building Council (USGBC) set out to explore these and other questions, through a detailed study of post-occupancy energy use in LEED-certified buildings. Out of 552 LEED for New Construction– (LEED-NC-) certified buildings invited to participate in the study, 121 provided actual records for at least one year of energy bills. Respondents included K–12 schools, libraries, laboratories, interpretive centers, offices, multiunit residential buildings, multiuse buildings, public order buildings, and others.

Building energy use was assessed in three ways:

- The Energy Use Index (EUI), which represents actual energy use in thousands of Btu per square foot
- Energy Star, which rates building energy use on a scale of 1 to 100, benchmarked against a national database of energy use in existing buildings
- Percent energy savings compared with ASHRAE 90.1, the baseline used for modeled energy savings in LEED (and in the estimates of energy benefits presented in this report).

The results of the study indicate that (1) LEED buildings achieve an average reduction in energy use of 25%, when compared with the median energy use of existing buildings documented in the Department of Energy's Commercial Buildings Energy Consumption Survey (CBECS), and (2) the level of savings increases as LEED level increases.[2] That is, LEED succeeds in one of its primary goals, which is to encourage energy-efficient building and reward higher levels of energy performance. On average, Gold and Platinum buildings use 45% less energy than the CBECS average—close to the first interim goal of a 50% energy reduction set by a not-for-profit coalition of national architectural, design, and green building organizations called the 2030 Challenge.[3] The average Energy Star score for LEED buildings was 68, indicating that LEED buildings are, on average, more efficient than 68% of existing buildings. Average reduction in energy use compared with ASHRAE 90.1 was 28%, compared with an average projected reduction of 25%.

Although average projected energy savings were realized across the set of buildings, individual building performance varied significantly from projections. Thirty percent of LEED buildings surveyed achieved greater savings than projected, and 25% used more energy than expected, with several experiencing significant energy problems and using more energy than the code baseline. Schools, libraries, and multiuse buildings tended to realize greater savings than projected, while laboratories used twice the energy projected. Buildings in warm to hot climates achieved minimal energy-use reductions when compared with CBECS, whereas buildings in cool or mixed climates achieved greater reductions.

The results of the NBI/USGBC study complement the energy data and benefits estimates developed in this report. They also add some important insights that should help identify topics for further research, and guide future revisions to LEED credits and processes. First, modeled energy use provides a good sense of the overall performance of LEED buildings, but is a poor predictor of energy use in individual buildings; more work is needed on refining and calibrating energy systems and modeling in green buildings—especially for high-energy-use buildings such as laboratories. Second, the ASHRAE 90.1 baseline used by LEED is not as aggressive as expected and is similar to the average energy use of nongreen buildings in CBECS. This observation, which is based on actual modeling results submitted to the USGBC, differs from the previous industry assumption that the ASHRAE 90.1 standard would create buildings significantly more efficient than average existing buildings. Further research will be needed to understand the reason for these different results and their implications, as the ASHRAE 90.1 standard and LEED requirements become more stringent over time. It is worth noting that the 2010 version of ASHRAE 90.1 is expected to be 20–35% lower energy use than the 2007 version which is itself about 10% lower energy use than the 2004 version.

APPENDIX J
ASSUMPTIONS USED FOR CALCULATIONS OF WATER SAVINGS

Data on water rates were not available for most building types in the data set, so in order to model expected savings in green buildings, we used assumptions based on recent surveys of water rates and water use. For water rates, we used the median water and wastewater rates for commercial buildings documented in the Raftelis 2006 National Water Rate Survey.[1] There is no national survey of water use and prices across building types, though several independent studies have been conducted to track and benchmark this information. Assumptions for baseline water use in different building types were based on a number of sources, including the 2001 Vickers and Associates "Handbook of Water Use and Conservation,"[2] the 2000 Aquacraft survey "Commercial and Industrial End-Uses of Water,"[3] an Aquacraft study on submetering in multiunit residential buildings (both Aquacraft surveys are used as reference sources in the industry),[4] and conversations with water engineers.[5] Since the water savings reported by buildings in the study data set do not generally include irrigation or process water, for the purposes of this model, we assumed (conservatively) that total reduction in water use is proportional to reported savings. We applied wastewater charges to indoor water usage only for those building types for which irrigation water use is often substantial.

Because of the small sample size and lack of differences in water savings between building types (see figure 1.16, "Water-Use Reductions by Building Type"), we use the median water-use reduction for the entire data set (35%) to model benefits across all building types. So, for example, typical water use in schools can be used to model the differences in savings by LEED level, based on median reductions in water use for buildings in the data set at each LEED level. Table J.1 shows the baseline water-use assumptions, data sources, percent reductions in water use, and first-year water costs that were used to model water benefits for different building types and LEED levels.

Water rates are increasing across the country, largely because of significant needs for infrastructure investments and repair. A national sampling of water infrastructure plans and water rates bears out the need for increased funding:

- The District of Columbia's sewer system currently overflows dozens of times each year into Rock Creek and the Anacostia and Potomac rivers, damaging ecosystems and restricting public use of area parks and waterways. The D.C. Water and Sewer Authority has embarked on a multiyear, $1.3 billion project to improve the control system for the district's wastewater and storm-water sewers, which will reportedly require 7% to 9% increases in water rates each year over the coming 10 to 40 years.[6]

Building type	Annual water use (gallons/sf)	Annual waste-water use (gallons/sf)	Data source	Percent reduction in water use	First-year cost savings ($/sf)
Schools	60	16	Aquacraft CIEW[1]	39%	$0.08
Offices	26	11	Aquacraft CIEW	39%	$0.04
Apartments	52	52	For a 1,000-square-foot unit, 143 gallons/day/unit, based on Aquacraft submetering study	39%	$0.11
Health care facilities[2]	55	55	Communication with Kim Shinn, TLC Engineering	39%	$0.12
Religious-assembly buildings	15	15	Based on average use reported by buildings in data set	39%	$0.03
Certified	60	16	Aquacraft (schools)	21%	$0.04
Silver	60	16	Aquacraft (schools)	36%	$0.07
Gold	60	16	Aquacraft (schools)	39%	$0.08
Platinum	60	16	Aquacraft (schools)	55%	$0.11

1 Benedykt Dziegielewski et al., "Commercial and Institutional End Uses of Water," American Water Works Association, 2000.
2 Assumes reductions in both Energy Policy Act of 1992 and process water.

- Boston water prices rose steadily from 2002 to 2007, and were projected to rise 8.5% in 2008.[7]
- An 11.5% increase in New York City rates was projected for 2008.[8]
- In Tucson, a switch from groundwater to river water as the major municipal source will require a rate increase of 6.2% in 2009, with greater increases expected in the future.[9]
- A $3.9 billion overhaul of the sewer and water systems in Atlanta will be partially financed through a 10% annual increase in water and sewer rates expected to continue for years.[10]

Older cities across the country are facing major overhauls of aging water systems requiring similar increases in customer water rates. Many younger cities face similar costs as they seek to increase water supplies in the face of rapidly expanding populations.

Table J.2 shows the cost of water-saving features in seven buildings in the data set.

Building/ Location	Total project cost	Year	Reported water-use reduction	Costs, savings, and net effect
Oregon Health Science University Center for Health and Healing, Portland, OR	$150,000,000	2006	61%	+$50,000 rainwater harvesting and recycling system +Bioreactor cost (financed through third-party vendor; did not create additional cost for owner) −$50,000 incentive grant from city =$0 additional initial cost to owner
Queens Botanical Garden, Queens, NY	$15,505,000	2007	41%	−$12,500 irrigation system savings −$2,000 waterless urinal plumbing savings +$25,000 additional cost for composting toilet =$10,500 net additional cost
Office of Emergency Management, Brooklyn, NY	$38,000,000	2006	33%	−$12,500 irrigation system savings −$1,000 waterless urinal plumbing savings =$13,500 cost savings
Brooklyn Children's Museum, Brooklyn, NY	$36,600,000	2007	30%	−$12,500 irrigation system savings −$2,500 waterless urinal plumbing savings =$15,000 cost savings
National Business Park 306, Annapolis Junction, MD	$24,130,863	2006	not available	+$30,500 efficient landscaping and fixtures
Washington-Lee High School, Arlington, VA	$95,400,000	2009	46%	Reduced irrigation, low-flow faucets and shower-heads, waterless urinals, and dual-flush toilets; savings from irrigation reduction offset small additional costs of fixtures =$0 additional cost

APPENDIX K
GREEN BUILDING SURVEY INSTRUMENT

1. To what degree were the following factors a motivation behind your community, congregation, or institution's decision to build green?

    - Financial
    - Faith-based
    - Other

2. If the decision was partly faith-based, how was the decision-making process to build green influenced by your faith community, congregation, or institution's theological beliefs, sacred texts, or religious values? Could you provide some examples?

3. Is the stewardship of God's creation, or care for the earth, or some similar language a part of your organization's general mission statement? If so, could we receive a copy of the mission statement?

4. Was there any news story or public statement made regarding the green building project that expressed your community's commitment to theological or ethical values regarding the environment? If so, are copies available?

5. Did you receive any support, consultation or encouragement from any national organizations? If so, what are the national organizations?

6. Was there any resistance within your congregation, organization, or institution to building green? If so, what were the core arguments for or against? How were conflicts resolved?

7. After completion of the green building project, what has been the community, congregation, or institution's reaction/response? Has it had an impact upon the faith life of the community in any appreciable or tangible ways? What has your community, congregation, or institution learned from the project? Have you received any special recognition from an outside agency, association or organization?

8. What was the additional cost incurred (the cost premium) to construct a green building vs. a conventional building?

9. What are the projected and/or actual energy savings?

10. What are the projected and/or actual water savings?

APPENDIX L
GLOBAL ASSUMPTIONS FOR PART IV

Assumption	Value	Source
CARBON DIOXIDE (CO_2) EMISSIONS		
U.S. building energy CO_2 emissions in 2005	2,556 tons	Energy Information Agency (EIA) Annual Energy Outlook 2009, Early Release (www.eia.doe.gov/oiaf/aeo/)
U.S. energy carbon-intensity projections		Calculated from EIA Annual Energy Outlook 2009, Early Release (www.eia.doe.gov/oiaf/aeo/pdf/overview.pdf)
BUILDING SECTOR ENERGY USE		
Estimated U.S. energy consumption in 2010 (quads)	99.9	EIA Annual Energy Outlook 2009, Early Release
Percentage of U.S. energy consumed by buildings in 2010	41%	EIA Annual Energy Outlook 2009, Early Release
Percentage of U.S. energy from embodied energy and industrial office space	4%	Good Energies, Inc. (GEI) estimate
BUILDING ENERGY USE		
Normalized energy use for existing buildings, 2010	100%	
Annual arithmetic reduction in energy use in buildings existing in 2010 from minor retrofits	0.20%	GEI estimate
Energy use of new buildings in 2010 as a percentage of existing buildings in 2010	85%	Mark Frankel, technical director, New Buildings Institute
Annual arithmetic reduction	0.50%	GEI estimate
Energy use of new green buildings in 2010 as a percentage of new conventional buildings in 2010	65%	GEI estimate, based on U.S. Green Building Council (USGBC) data
Annual arithmetic reduction	1.50%	GEI estimate
Energy use of conventionally retrofitted buildings in 2019 as a percentage of existing buildings in 2010	80%	Steve Nadel, executive director, American Council for an Energy-Efficient Economy
Annual arithmetic reduction	0.50%	GEI estimate

Assumption	Value	Source
Energy use of a green retrofitted building as a percentage of a conventional retrofit	75%	GEI estimate
Annual arithmetic reduction	1.50%	GEI estimate

BUILDING STOCK

Square feet built new and demolished annually	Variable	EIA projections
Percentage of new green construction in 2010	5%	Based on USGBC LEED registrations and certifications, and McGraw-Hill's *Green Outlook 2009*
Percentage of comprehensive green retrofits in 2010	0.50%	GEI estimate

RENEWABLE ENERGY

Average percentage of green building electricity from on-site renewables in 2010	2.00%	Indicative data from the USGBC
Average percentage of green building electricity from off-site renewables in 2010	12%	Indicative data from the USGBC
Factor by which impact of off-site renewables is discounted	50%	
Percentage of building energy attributable to electricity	74%	EIA Annual Energy Outlook 2009, Early Release
Average percentage of primary energy lost in generation of electricity in U.S.	65%	EIA Annual Energy Databook, 2009, Early Release; U.S. Department of Energy (DOE), Annual Energy Review, 2007
Average percentage loss of electricity from transmission and distribution	9%	DOE, Annual Energy Review, 2007 (www.trb.org/news/blurb_detail.asp?id=9196)
Off-site renewable-energy capacity factor	30%	Based on capacity factor for wind, solar, thermal, and hydro
On-site renewable-energy capacity factor	20%	Based on capacity factor for photovoltaics

COSTS AND BENEFITS

Additional costs to build green (/sf)	($3.87)	GEI data set (median)

ABOUT THE AUTHORS

GREGORY "GREG" KATS is senior director and director of climate change policy at Good Energies, a global private investor in clean-energy technologies. He leads Good Energies' investments in energy efficiency and green buildings. He is also a senior advisor and member of the investment committee of Osmosis Capital LLP, a London-based private equity fund of funds operating in the low carbon sector.

Previously, Kats served as the director of financing for energy efficiency and renewable energy at the U.S. Department of Energy, the country's largest clean-technology development and deployment program. He was the founding chair of the International Performance Measurement & Verification Protocol, which has served as a technical basis for $10 billion in building upgrades and been translated into ten languages.

Kats is a founder of the American Council on Renewable Energy, and the New Resource Bank. He is founding Chair of the Energy and Atmosphere Technical Advisory Group for LEED, and was the principal advisor in developing Green Communities, the national green affordable-housing design standard. Earlier in his career, Kats held senior marketing and management positions at Reuters, in Paris, Geneva, and London. He served on the Parliamentary Committee on Environmental Sustainability in the United Kingdom's House of Lords.

Kats earned an MBA from Stanford University and, concurrently, an MPA from Princeton University. He was a Morehead Scholar at UNC and is a certified energy manger. He was a principal author of *Green Office Buildings: A Practical Guide to Development* (ULI, 2005). Kats serves on numerous boards and as sustainability advisor to CalPERS. He is a frequent keynote speaker at national clean-energy technology, venture capital, and real estate conferences.

JON BRAMAN is a researcher with Good Energies and a consultant with Bright Power, Inc., a New York-based renewable-energy and energy-efficiency firm where he focuses on implementing energy efficiency in affordable multifamily housing. Braman was a contributing researcher to the 2006 report "Greening America's Schools: Costs and Benefits." He holds a B.S. in ecology and evolutionary biology from Yale University.

MICHAEL JAMES serves as executive director of the Center for Catholic Education at the Lynch School of Education at Boston College. James has over 20 years of experience in Catholic education administration, and previously served as vice president of the Association of Catholic Colleges and Universities, and in a range of administrative and academic roles at the University of Notre Dame, Indiana University, and Mount Marty College.

STEVEN I. APFELBAUM, principal and founder of Applied Ecological Services, is one of the leading ecological consultants in the United States, providing technical restoration advice and win-win solutions where ecological and land development conflicts arise.

DANA BOURLAND is senior director of Green Communities for Enterprise Community Partners, a national nonprofit organization. Bourland directs all aspects of Green Communities; coleads environmental strategy for the organization; and serves as managing director of the Green Communities Offset Fund. She previously worked at the Maryland Department of Planning, where she implemented smart growth–related policies. A former Peace Corps Volunteer, Bourland has a master's degree in planning from the University of Minnesota.

TOM DARDEN is chief executive officer of Cherokee Investment Partners, the leading private equity firm investing in brownfield redevelopment; Cherokee has 525 properties and more than $2 billion under management worldwide.

JILL C. ENZ is an ecological designer and project manager with Applied Ecological Services. She is responsible for the design and management of conservation development and large-scale land planning projects.

DOUGLAS FARR is president and chief executive officer of Farr Associates, a planning and architecture firm based in Chicago. An architect and urban designer, he has served as cochair of the Environmental Task Force of the Congress for the New Urbanism, chair of the American Institute of Architects' Chicago Committee on the Environment, and chair of the U.S. Green Building Council's Leadership in Energy and Environmental Design (LEED) for Neighborhood Development Core Committee.

ROBERT F. FOX JR. is a partner in Cook + Fox Architects and a founder of Terrapin Bright Green, a consulting and strategic planning company. He is a member of New York mayor Michael Bloomberg's Advisory Committee on Sustainability and Long-Term Planning and has received numerous awards, including a leadership award from the U.S. Green Building Council and the Urban Visionary Award from the Cooper Union.

LAWRENCE FRANK holds the J. Armand Bombardier Chair in Sustainable Transportation at the University of British Columbia, and is a senior fellow at the Brookings Institution's Metropolitan Policy Program. Frank led a five-year, $4.6 million research program known as SMARTRAQ (Strategies for Metropolitan Atlanta's Regional Transportation and Air Quality), which tested the impacts of land use and transportation decisions on travel choices, physical activity, and air quality.

ROBIN GUENTHER, FAIA, is principal of Perkins+Will, in New York, and is involved in a wide range of advocacy initiatives at the intersection of health care architecture and sustainable policy. In 2004, she was awarded the Changemaker Award from the Center for Health Designs. She co-coordinates the Green Guide for Health Care, and co-authored "Sustainable Healthcare Architecture" with Gail Vittori in 2007. She is a board member of Healthy Building Network, the Center for Health Design, and Practice Greenhealth.

ADELE HOUGHTON, AIA, is principal of Adele Houghton Consulting, a sustainability consulting company specializing in solutions for the design and construction industry, developers, nonprofits, and government agencies. From 2005 to 2008, Houghton served as project manager and pilot-project coordinator for the Green Guide for Health Care, a voluntary, best-practices building tool kit tailored to the health care environment.

SARAH KAVAGE is a Seattle-based artist and writer with a master's degree in urban planning from the University of Washington. As special projects manager at Lawrence Frank & Company, she focuses on developing evidence-based tools to support decision making and policy.

JOHN (SKIP) LAITNER is economic analysis director at the American Council for an Energy-Efficient Economy. He has more than 30 years of experience studying the economics of energy and energy efficiency, including the assessment of cost-effective technologies that can reduce both energy use and related greenhouse-gas emissions.

JOE LEHMAN is a doctoral student in the higher education administration program at Boston College; a Catholic priest; and a member of the Franciscans, T.O.R.

CHRISTOPHER B. LEINBERGER is a visiting fellow at the Brookings Institution, where he researches and develops strategies to create walkable urban places and management models for metropolitan areas. Leinberger has written award-winning articles for publications including the *Atlantic Monthly*, the *Wall Street Journal*, and *Urban Land*, and has served as author or contributor for eight books.

GARY JAY SAULSON is director of corporate real estate for the PNC Financial Services Group. He is responsible for all of PNC's non-lending real state functions, and leads PNC's environmental strategy. Under his leadership, PNC has had more buildings certified by the U.S. Green Building Council than any other company in the world.

CRAIG Q. TUTTLE is a senior landscape architect and planner with Applied Ecological Services, where he serves as a lead designer on conservation development projects throughout the Midwest; the projects range in size from 40-acre subdivisions to 1,000-acre site-planning projects.

GAIL VITTORI is codirector of the Center for Maximum Potential Building Systems, an Austin-based nonprofit sustainable planning and design firm, established in 1975, that is engaged with private and public sector sustainable development, and design and materials projects. Ms. Vittori is the 2009 chair of the U.S. Green Building Council's board of directors and was a Loeb Fellow at Harvard University's Graduate School of Design. She is the founding chair of the Leadership in Energy and Environmental Design for Healthcare core committee, and is convener and co-coordinator of the Green Guide for Health Care.

STOCKTON WILLIAMS is director of green economy initiatives at Living Cities. Williams previously served as senior vice president and chief strategy officer at Enterprise Community Partners, a leading national nonprofit that creates affordable homes and economic development opportunities in low-income communities across the United States. Williams was instrumental in the design and development of Green Communities, Enterprise's groundbreaking commitment to mainstream sustainability in the affordable-housing industry. As of 2009, Green Communities has served as the design standard for 15,000 units of green affordable housing.

SALLY WILSON is senior vice president at CB Richard Ellis and is the firm's global director of environmental strategy—managing strategy, implementation, and communications related to global environmental initiatives. CB Richard Ellis is the world's largest commercial real estate services firm, and was named one of the 50 "best in class" companies by *BusinessWeek*, and one of the 100 fastest-growing companies by *Fortune* in 2007.

NOTES

PART I

1. Robert Charles Lesser & Co. (RCLCO), "Measuring the Market for Green Residential Development," January 2008 (www.usgbcncr.org/Documents/MarketforGreenResidentialDevelopment.pdf). See also "Perspective: Measuring Consumer Demand for Green Homes," in this volume.

2. More than half of the approximately 350 buildings originally contacted for this study were not able to provide cost-premium estimates, and were therefore excluded from the final data set.

3. We chose LEED as the green standard for the study because it is the most widely used green-building rating system in the world; many state and local governments, corporations, and institutions have adopted it as a standard for future construction. LEED is an evolving standard administered through an open system of committee and public review, and shares the large majority of its requirements with other international green-building standards (see section 2.9, "International Green Building"). Roughly 15% of the buildings included in the study were designed to another green building standard, often for a specific building type; these include the Massachusetts Collaborative for High-Performance Schools, the Enterprise Green Communities standard, and the Green Guide for Health Care. Where possible, we have used our knowledge of the other standards' requirements to assign an equivalent LEED level to non-LEED buildings. If no LEED equivalent level could be assigned, the building was excluded from impacts calculated on the basis of LEED level.

4. Throughout this book, references to the "study data set" denote the information in Appendix C.

5. To maintain consistency in baselines, only the 155 U.S. buildings in the data set were used for averages and in the cost-benefit models. The 11 international buildings are generally evaluated and discussed separately.

6. In fact, LEED NC does offer a point for post-occupancy monitoring of one aspect of indoor environmental quality: thermal comfort. As of December 2007, roughly 50% of LEED-certified buildings had been awarded this point. However, LEED allows buildings to take a variety of approaches to satisfy this credit, so consistent data on indoor environmental impacts remains sparse among green buildings.

7. McGraw-Hill Construction, *Global Green Building Trends* (McGraw-Hill Construction, 2008).

8. Nick Berg, personal communication with author, 2007.

9. Michelle Rosenberger, GGLO Architects, personal communication with author, 2007.

10. World Business Council for Sustainable Development (WBCSD), "Energy Efficiency in Buildings: Business Realities and Opportunities," 2007 (www.wbcsd.org/DocRoot/lKDpFci8xSi63cZ5AGxQ/EEB-Facts-and-trends.pdf).

11. RCLCO, "Measuring the Market Demand for Green Residential Development," RCLCO, 2008 (www.rclco.com).

12. The data set consisted of mostly new construction, but also included some retrofits.

13. One building in the data set reported a "negative premium": that is, lower costs than would have been incurred with conventional design.

14. Roughly one-quarter of the buildings in the data set reported receiving public incentives or grants to support the green features. Incentives and grants reduced the median green premium for the data set to 1.4%, and the mean green premium to 2.4%. In most cases, however, incentives do not significantly alter the cost-effectiveness of green design. As the owner of one building in the data set reported, grants "were appreciated, but not a deciding factor."

15. We used medians more often than means in this study because they provide a sense of typical value without being skewed by isolated extreme data points. Cost and benefit estimates are generally rounded to the nearest whole dollar amount. Uncertainties about the data, including future price escalation, make greater precision misleading.

16. Steven Winter Associates, Inc., "LEED Cost Study: Final Report," October 2004 (www.wbdg.org/ccb/GSAMAN/gsaleed.pdf).

17. Ibid.

18. Davis Langdon, "Cost of Green Revisited: Reexamining the Feasibility and Cost Impact of Sustainable Design in the Light of Increased Market Adoption," 2007 (www.davislangdon.com /upload/images/publications/USA/The%20Cost%20of%20Green%20Revisited.pdf).

19. Ibid.

20. Kevin Hall, OWP/P, personal communication with author.

21. Green community design results in additional energy savings from reduced driving. See part 2, "Costs and Benefits of Green Community Design."

22. Most of the reported energy-use reductions in the data set are based on a comparison of modeled energy use for the green building with modeled energy use for a conventionally constructed version of the same building that meets the minimum efficiency requirements of ASHRAE 90.1 1999. (Appendix H discusses considerations in choosing the baselines and standards for energy-savings comparisons.) LEED Energy and Atmosphere Credit 1 requires buildings to report energy-cost reductions from an ASHRAE 90.1 baseline model. Published information on green buildings often cites energy savings without specifying whether the savings calculations are based on cost or usage. The survey we used asked about reductions in energy use, but reported reductions in energy use might be expected to include some figures calculated on the basis of cost. For a given building, the difference between reductions in energy cost and reductions in usage will depend on rate structures and the different costs of energy by fuel type. Cost and usage savings may be significantly different for individual buildings, but are not considered significant for the overall dollar- and emissions-savings estimates presented here, which are based on a relatively large portfolio of buildings.

23. See U.S. Department of Energy (DOE), Energy Efficiency and Renewable Energy, "2007 Buildings Energy Data Book" (www.btscoredatabook.net/docs%5CDataBooks%5C2007_BEDB.pdf).

24. New Buildings Institute, "Energy Performance of LEED for New Construction Buildings," 2007 (www.newbuildings.org/downloads/Energy_Performance_of_LEED-NC_Buildings-Final_3-4-08b.pdf).

25. This change was initiated by LEED's Energy and Atmosphere Technical Advisory Group, chaired by Greg Kats. See www.usgbc.org.

26. Energy Information Administration (EIA), "Commercial Buildings Energy Consumption Survey" (CBECS; www.eia.doe.gov/emeu/cbecs/contents.html).

27. Except for multifamily building expenditures, baseline energy expenditures by building type are from EIA, "Energy Consumption Survey," 2003 (www.eia.doe.gov/emeu/cbecs/); multifamily building expenditures are from EIA, "Residential Energy Consumption Survey" (www.eia.doe.gov /emeu/recs/contents.html). For release dates of the ongoing 2007/2008 CBECS, see www.eia.doe .gov/emeu/cbecs/. Baseline expenditure data were adjusted to 2008, assuming an annual increase in energy prices of 3% per year from 2003 to 2008. (This is lower than the actual increase in energy prices increase over that period, to account for a general improvement in the efficiency of conventional buildings over the same period. Given the increase in plug loads from computers and other electronics, it is unclear whether the efficiency increases have really decreased energy use.)

28. Appendix H includes a description of energy expenditures and savings data and sources for figures 1.10 and 1.11.

29. See EIA, "Total Electric Power Summary Statistics" (www.eia.doe.gov/cneaf/electricity/epm /tablees1b.html).

30. See EIA, "U.S. Price of Natural Gas Sold to Commercial Consumers" (http://tonto.eia.doe.gov /dnav/ng/hist/n3020us3a.htm).

31. See EIA, "Total Energy Expenditures by Major Fuel for All Buildings, 2003" (www.eia.doe.gov /emeu/cbecs/cbecs2003/detailed_tables_2003/2003set14/2003html/c2a.html). Non-electricity fuels include natural gas (16% of total energy expenditures), heating oil (7%), and district heat (2%). For the purposes of this calculation, we assumed that natural gas supplied all non-electricity energy.

32. There is a cost associated with uncertainty about future energy prices—a cost that can be reduced through green, energy-efficient design. This book does not estimate the financial value of the reduction in risk associated with future energy costs, but the value is material for many institutions. For a thorough review of the cost of uncertainty about energy prices, see Amory Lovins, *Small Is Profitable* (Snowmass, Colo.: Rocky Mountain Institute, 2006).

33. Cathy Turner and Mark Frankel, "Energy Performance of LEED-NC Buildings," 2008 (www.newbuildings.org/downloads/Energy_Performance_of_LEED-NC_Buildings-Final_3-4-08b.pdf).

34. Lori Bird and Blair Swezey, "Green Power Marketing in the United States: A Status Report," 6th ed. (www.eere.energy.gov/greenpower/resources/pdfs/35119.pdf).

35. For a detailed discussion of the economic benefits of reducing the use and generation of conventional energy, see Lovins, *Small Is Profitable*. See also www.smallisprofitable.org/.

36. McKinsey & Company, "Reducing U.S. Greenhouse Gas Emissions: How Much at What Cost?" March 14, 2008 (www.mckinsey.com/clientservice/ccsi/greenhousegas.asp).

37. Ryan Wiser, Mark Bolinger, and Matt St. Clair, "Easing the Natural Gas Crisis: Reducing Natural Gas Prices through Increased Deployment of Renewable Energy and Energy Efficiency," January 2005, 40 (http://eetd.lbl.gov/ea/ems/reports/56756.pdf).

38. Platts Research & Consulting, "Hedging Energy Price Risk with Renewables and Energy Efficiency," September 2004.

39. Massachusetts Division of Energy Resources, "2002 Energy Efficiency Activities," Summer 2004 (www.mass.gov/Eoeea/docs/doer/electric_deregulation/ee02.pdf).

40. WBCSD, "Energy Efficiency in Buildings," 2007.

41. Alex Wilson, "Cement and Concrete: Environmental Considerations," *Environmental Building News* (March 1993); available at http://www.buildinggreen.com/auth/article.cfm?fileName =020201b.xml; Ken Humphreys and Maha Mahasenan, "Toward a Sustainable Cement Industry," Substudy 8: Climate Change (Battelle, March 2002). Roughly one-half of CO_2 emissions are byproducts of cement manufacture and are not related to energy use. The Portland Cement Association estimates that for an average U.S. home, the amount of CO_2 emitted from cement production alone is equal to the amount of CO_2 emitted through the generation of 1.5 years of energy use in the home. See Portland Cement Association, *PCA Annual Yearbook* (2007).

42. Seppo Junnila, Arpad Horvath, and Angela Acree Guggemos, "Life-Cycle Assessment of Office Buildings in Europe and the United States," *Journal of Infrastructure Systems* 12, no. 1 (2006).

43. Environment Australia, Department of Environment and Heritage, "Greening the Building Life-Cycle" (http://buildlca.rmit.edu.au/CaseStud/EE/EEcommerc.html#location-2).

44. See, for example, Calstar cement: www.calstarcement.com/.

45. See the LEED reference guide: www.usgbc.org.

46. U.S. Green Building Council (USGBC), LEED credit tally.

47. See the Web site of the Regional Greenhouse Gas Initiative (www.rggi.org/) and the Web site of the Chicago Climate Exchange (www.chicagoclimateexchange.com). For a brief discussion of strengths and weaknesses of the Chicago Climate Exchange, see Abrahm Lustgarten, "CCX's New Competition," *Fortune*, September 1, 2006; available at http://money.cnn.com/2006/08/30 /news/economy/carbonexchange.fortune/).

48. See the Web site of the European Climate Exchange (www.europeanclimateexchange.com/) and www.pointcarbon.com.

49. See Jason Moresco, "Carbon Costs under Obama Cap and Trade," *Red Herring*, March 2, 2009 (www.redherring.com/Home/25894); Stephen Power, "Carbon Trading to Raise Consumer Energy Prices" *Wall Street Journal*, February 27, 2009 (http://online.wsj.com/article /SB123566843777484625.html#articleTabs%3Darticle).

50. The details of trading systems for carbon emissions will affect how this value accrues. For instance, end users of energy (such as building owners) should be awarded the CO_2 reduction credits and capture the value of reductions from building efficiency and renewable-energy investments they make in their buildings. It is unlikely, however, that the full value of CO_2 reductions in buildings will pass through to building owners; other mechanisms, such as set-asides for building efficiency, may be needed. Organizations that hold or represent large building portfolios may be able to aggregate building-efficiency gains across many buildings, and put them up for sale as offsets on a carbon-trading market. If the value of CO_2 reductions accrues only to utilities, then the value for individual green-building owners or investors will be minimal.

51. Intergovernmental Panel on Climate Change, "Summary for Policy Makers" (www.ipcc.ch/pdf /assessment-report/ar4/syr/ar4_syr_spm.pdf).

52. Stern Review on the Economics of Climate Change (http://www.hm-treasury.gov.uk/stern _review_report.htm).

53. See the EPA criteria pollutants page: www.epa.gov/air/urbanair/.

54. See http://epa.gov/air/urbanair/nox/hlth.html.

55. See www.epa.gov/oar/particlepollution/index.html and www.nrdc.org/air/pollution/qbreath.asp.

56. See www.epa.gov/camr/.

57. Michael Hopkins, "Emissions Trading: The Carbon Game," *Nature* 432 (November 2004):

268–270; Richard Clarkson and Katherine Deyes, *Estimating the Social Cost of Carbon*, Government Economic Service Working Paper 140 (HM Department of Treasury, 2002).

58. See Environmental Benefits and Mapping Program, a tool for estimating the benefits of emissions reductions (www.epa.gov/air/benmap/). Greg Kats, "Costs and Financial Benefits of Green Buildings," Capital E, 2003, also includes the benefits of emissions reductions in green buildings; see www.cap-e.com/ewebeditpro/items/O59F12807.pdf.

59. The energy-savings estimates in table 1.1 do not include reductions in the use of conventional power resulting from green power purchases; however, green power purchases are included in the CO_2 reduction scenarios in part 4.

60. DOE, "Geothermal Heat Pumps" (http://apps1.eere.energy.gov/consumer/your_home/space _heating_cooling/index.cfm/mytopic=12640).

61. Ibid.

62. Good Energies, Inc. (GEI) is an investor in Sage Glass, the leading electrochromic glass company; see www.sage-ec.com/.

63. Lawrence Berkeley National Laboratory (LBNL), "Advancement of Electrochromic Windows," 2006 (www.lbl.gov/Science-Articles/Archive/sabl/2007/Jan/Advance-EC-Windows.pdf).

64. Lou Podbelski, Sage Electrochromics, personal communication with author, 2007; see www.sage-ec.com.

65. The Database of State Incentives for Renewable Energy (DSIRE) is an online source of information on state, local, utility, and federal standards, incentives, and programs relating to renewable energy; see www.dsireusa.org/.

66. Amy Vickers, Vickers and Associates, personal communication with author, 2007.

67. Cathy Turner, "LEED Building Performance in the Cascadia Region: A Post-Occupancy Evaluation Report," Cascadia Region Green Building Council, 2006.

68. Ibid.

69. Kim Shinn, TLC Engineering, personal communication with author, 2007.

70. USGBC, LEED-NC credit tally, June 2007.

71. D. Langdon, "Cost of Green Revisited: Reexamining the Feasibility and Cost Impact of Sustainable Design in the Light of Increased Market Adoption," 2007 (www.greenerbuildings.com/resources /resource/cost-green-revisited-reexamining-feasibility-and-cost-impact-sustainable-design-1).

72. Peiffer Brandt, Raftelis Financial Consultants, personal communication with author, 2007.

73. Edwin H. Clark II, "Water Prices Rising Worldwide," 2007 (www.earth-policy.org/Updates /2007/Update64.htm).

74. EPA, Office of Water, "Clean Water and Drinking Water Infrastructure Gap Analysis," 2006.

75. U.S. Census Bureau, "Construction Grants Program and CWSRF Expenditures," 2002.

76. New York City Department of Environmental Protection, *Water Conservation Program* (Flushing, N.Y., 2006). Thanks to Warren Liebold, Director, Technical Services/Conservation Bureau of Customer Services, New York City Department of Environmental Protection.

77. University of Michigan, Center for Sustainable Systems, "U.S. Water Supply and Distribution: Fact Sheet," 2005; R. Myhre, "Water & Sustainability (Volume 3): U.S. Water Consumption for Power Production—The Next Half Century" 2002 (http://mydocs.epri.com/docs/public /000000000001006786.pdf).

78. Ronnie Cohen, Barry Nelson, and Gary Wolff, "Energy Down the Drain: The Hidden Costs of California's Water Supply," Natural Resources Defense Council, Pacific Institute, 2004 (www.nrdc.org/water/conservation/edrain/edrain.pdf).

79. USGBC, LEED-NC credit tally, December 2007.

80. Barbara Deutsch et al., "Re-Greening Washington, DC: A Green Roof Vision Based on Quantifying Storm Water and Air Quality Benefits," Casey Trees Endowment Fund and Limno-Tech, Inc., 2005 (www.greenroofs.org/resources/greenroofvisionfordc.pdf).

81. K. Acks, "Green Roofs in the NY Metropolitan Area: A Framework for Cost-Benefit Analysis of Green Roofs—Initial Estimates," Columbia University Center for Climate Systems Research, National Aeronautics and Space Administration, Goddard Institute for Space Studies, 2006 (http://ccsr.columbia.edu/cig/greenroofs/Green_Roof_Cost_Benefit_Analysis.pdf).

82. For a full list of partners, see www.greencommunitiesonline.org. Greg Kats, lead author of this report, was the principal advisor in developing this standard.

83. New Resource Bank (www.newresourcebank.com) has recently begun offering lower mortgage rates for green buildings.

84. William Bradshaw et al., "The Costs and Benefits of Green Affordable Housing," New Ecology and the Tellus Institute, 2005 (www.dnr.state.md.us/ed/finalcbreport.pdf).

85. National Energy Assistance Directors' Association, 2005 National Energy Assistance Survey (Washington, D.C.: National Energy Assistance Directors' Association, 2005), i–iv.

86. Bradshaw et al., "Green Affordable Housing," 2005.

87. Ernie Hood, "Dwelling Disparities: How Poor Housing Leads to Poor Health," Environmental Health Perspectives (May 2005).

88. See Simon Fraser University, "Breathe Easy Homes Reduce Asthma Symptoms for Children and Win International Award for FHS Researcher" (www.fhs.sfu.ca/news/featured-article/breathe-easy-homes-reduce-asthma-symptoms-for). Healthy home design in "breathe-easy" homes was accompanied by education on healthy cleaning practices.

89. American Lung Association, Epidemiology and Statistics Unit, "Trends in Asthma Morbidity and Mortality," 2006 (www.lungusa.org/atf/cf/%7b7a8d42c2-fcca-4604-8ade-7f5d5e762256%7d/asthma06final.pdf); see also www.pulmicortrespules.com/childhood-asthma/children-toddlers.aspx.

90. Barbara J. Lipman, "A Heavy Load: The Combined Housing and Transportation Burdens of Working Families," Center for Housing Policy, 2006 (www.nhc.org/pdf/pub_heavy_load_10_06.pdf).

91. Barbara McCann and Reid Ewing, "Measuring the Health Effects of Sprawl: A National Analysis of Physical Activity, Obesity, Chronic Disease," Smart Growth America and the Surface Transportation Policy Project, 2003 (http://www.smartgrowthamerica.org/report/HealthSprawl8.03.pdf).

92. Dan W. Reicher, Director, Climate Change and Energy Initiative, Google.org, testimony before the Senate Committee on Finance, February 27, 2007.

93. Bradshaw et al., "Green Affordable Housing," 2005.

94. Neil E. Klepeis et al., "The National Human Activity Pattern Survey," LBNL, 2001.

95. See LBNL, "Indoor Air Quality Scientific Findings Resource Bank," for summaries and links to major research on the impact of indoor air quality on health and productivity (http://eetd.lbl.gov/ied/sfrb/). W. J. Fisk, "Health and Productivity Gains from Better Indoor Environments and Their Implications for the U.S. Department of Energy," 1999 (available at www.rand.org/scitech/stpi/Evision/Supplement/fisk.pdf); www.healthyhousing.org/clearinghouse/docs/Article0091.pdf; R. Ulrich, "View through a Window May Influence Recovery from Surgery," Science 224 (April 1984); Hood, "Dwelling Disparities."

96. Fisk, "Health and Productivity Gains"; Carnegie Mellon University, Center for Building Performance Diagnostics, "Building Investment Decision Support [BIDS] Tool" (http://cbpd.arc.cmu.edu/bids/).

97. USGBC Research Committee, "A National Green Building Research Agenda," 2007 (www.usgbc.org/ShowFile.aspx?DocumentID=3402).

98. Earlier versions of LEED devoted over one-fifth of possible points to IEQ credits.

99. Bob Thompson, EPA Indoor Environmental Management branch chief, personal communication with author, 2008.

100. Carnegie Mellon University, Center for Building Performance Diagnostics, BIDS Tool; 2007 summary data provided by Vivian Loftness.

101. LBNL, "Indoor Air Quality Scientific Resource Bank" (www.iaqscience.lbl.gov).

102. Don Aumann et al., "Windows and Classrooms: A Study of Student Performance and the Indoor Environment," California Energy Commission, 2004 (http://www.eceee.org/conference_proceedings/ACEEE_buildings/2004/Panel_7/p7_1/).

103. Ulrich, "Recovery from Surgery," 1984.

104. Sean Candrilli and Josephine Mauskopf, "How Much Does a Hospital Stay Cost?" RTI Health Solutions (www.rtihs.org/request/index.cfm?fuseaction=display&PID=6465).

105. Peter Boyce, Claudia Hunter, and Owen Howlett, "The Benefits of Daylight through Windows," Rensselaer Polytechnic Institute, Lighting Research Center, 2003 (www.lrc.rpi.edu/programs/daylighting/pdf/DaylightBenefits.pdf).

106. Dave Wood, Sidwell Friends Middle School, personal communication with author, 2007.

107. Greg Kats, "Greening America's Schools: Costs and Benefits," 2006 (www.buildgreenschools.org/documents/pub_Greening_Americas_Schools.pdf); correspondence with Moseley Architects, 2005.

108. S. Abbaszadeh et al., "Occupant Satisfaction with Indoor Environmental Quality in Green Build-

ings," *Proceedings of Healthy Buildings,* vol. 3 (2006): 365–370; available at http://repositories
.cdlib.org/cedr/cbe/ieq/Abbaszadeh2006_HB/.

109. Ibid.

110. Paladino and Company, Inc., "Performance Evaluation Report, PNC Green Branch Program,"
2007. Also see "Perspective: Birth of the Green Branch Bank."

111. Jacqueline Vischer, *Workspace Strategies: Environment as a Tool for Work* (Chapman and Hall,
1996).

112. Turner and Frankel, "Energy Performance," 2008.

113. National Research Council, *Green Schools: Attributes for Health and Learning* (Washington,
D.C.: National Academies Press, 2006).

114. For other recent estimates of the value of health and productivity improvements in green buildings,
see www.lincolnescott.com/refresh and Kats, "Green Buildings."

115. The U.S. General Services Administration's standard for rentable office space is 230 square feet
per person; see GSA, Office of Government Policy, "Real Property Performance Results," 2002.
The figure of $3,000 per person is significantly below per capita health spending in the United
States, which was estimated at over $7,000 in 2008 (http://content.healthaffairs.org/cgi/content
/abstract/27/1/14). A lower baseline for health costs is used to account for the portion of health
costs typically covered by employers. See "Facts on the Cost of Health Insurance and Health
Care" (www.nchc.org/facts/cost.shtml).

116. Fisk, "Health and Productivity Gains."

117. Rachel Dewane and Breeze Glazer contributed to the research for this section.

118. U.S. Department of Commerce, International Trade Administration, "Health Care Services
Sector, 2007" (www.trade.gov/investamerica/health_care.asp).

119. EIA, CBECS, 2003 (www.eia.doe.gov/emeu/cbecs/). Each year, CBECS reports on 8,000
inpatient health care buildings and on 121,000 outpatient health care buildings.

120. CBECS, 2003.

121. Brendan B. Read, "Analysis: President Obama's New Green Stimulus Law," February 18, 2009
(http://green.tmcnet.com/topics/green/articles/50701-analysis-president-obamas-new-green
-stimulus-law.htm); www.green.tmcnet.com.

122. Karen Ehrhardt-Martinez and Skip Laitner, "The Size of the U.S. Energy Efficiency Market:
Generating a More Complete Picture," American Council for an Energy-Efficient Economy
(ACEEE), 2008 (www.aceee.org/pubs/e083.htm).

123. Sarah White and Jason Walsh, "Greener Pathways: Jobs and Workforce Development in the
Clean Energy Economy," Center on Wisconsin Strategy, 2008 (www.greenforall.org/resources
/greener-pathways-jobs-and-workforce-development-in).

124. Although macroeconomic models show small utility-job losses from energy efficiency and reduced
power sales, in reality these losses are more likely to show up as shifts in the utility business: from
selling fossil-fuel-based energy to selling renewable energy, or establishing programs or business
models that are based on energy efficiency. For instance, utilities often perform energy-efficiency
upgrades, and thus gain some of the job increases that result from overall efficiency improvements.

125. John (Skip) Laitner, R. Neal Elliot, and Maggie Eldridge, "The Economic Benefits of an Energy
Efficiency and Onsite Renewables Strategy to Meet Growing Electricity Needs in Texas,"
ACEEE, 2007 (www.aceee.org/pubs/e076.htm). Job-creation studies typically report results
in job-years: one year of full-time employment for one person.

126. Black and Veatch, "Economic Impact of Renewable Energy in Pennsylvania," Heinz Endowments, Community Foundation for the Alleghenies, 2004 (www.cleanenergystates.org/library/pa
/PA%20RPS%20Final%20Report.pdf).

127. Joanne Wade, Victoria Wiltshire, and Ivan Scrase, "National and Local Employment Impacts of
Energy-Efficiency Investment Programmes," Association for the Conservation of Energy, 2000
(www.ukace.org/publications/ACE%20Research%20(2000-04)%20-%20National%20and
%20Local%20Employment%20Impacts%20of%20Energy%20Efficiency%20Investment
%20Programmes%20%5BVolume%201%20Summary%20Report%5D).

128. $3.50/sf represents a typical green premium for office buildings in the book data set.

129. At the same time, however, the occupant or building owner must pay back the borrowed funds,
which slightly reduces the increased spending capacity derived from energy savings.

130. EPA, "Waste Characterization Reports," 2003 (www.epa.gov/osw/nonhaz/municipal/pubs

/msw03rpt.pdf); correspondence with Kim Cochran, EPA.

131. USGBC, LEED-NC credit tally, 2007.

132. Assumed municipal solid-waste generation and diversion rate from EPA, Office of Solid Waste, "Municipal Solid Waste in the United States: Facts and Figures," 2003 (www.epa.gov/garbage /pubs/msw2001.pdf). Note that in 2007, municipal solid waste, including C&D waste, was estimated at roughly 250 million tons, so a 50-million-ton reduction in C&D waste would be a roughly 10% reduction in total U.S. waste.

133. California Integrated Waste Management Board (CIWMB), "Diversion Is Good for the Economy: Highlights from Two Independent Studies on the Economic Impacts of Diversion in California," CIWMB, 2003 (www.p2pays.org/ref/35/34533.pdf).

134. R. W. Beck, Inc., "California Recycling Economic Information Study," CIWMB, 2001 (www.ciwmb.ca.gov/Agendas/MtgDocs/2002/01/00007124.pdf).

135. CIWMB, "Diversion in California," 2003.

136. Omar Freilla, executive director, Green Worker Cooperatives, personal communication with author, 2007.

137. McGraw-Hill Construction, *Green Outlook 2009* (McGraw-Hill Construction, 2008). See also H. Bernstein, "Green Building SmartMarket Report," McGraw-Hill Industry Analytics, 2006 (www.construction.com/SmartMarket/greenbuilding/default.asp).

138. J. Spivey, "Commercial Real Estate and the Environment," CoStar, 2008 (www.costar.com/news/Article.aspx?id=D968F1E0DCF73712B03A099E0E99C679).

139. Ernst & Young, "Real Estate Market Outlook 2007," Ernst & Young Global Unlimited, 2007.

140. Theddi Chappell and Dan Kohlhepp, "Case Study: One and Two Potomac Yard," Pacific Security Capital, 2007 (http://www.wbdg.org/references/cs_potomac.php).

141. David Cohen, Fireman's Fund Insurance, personal communication with author, 2007.

142. E. Rand, "A Spec of Green: Award Winner Does Well by Doing Good," Development Online, National Association of Industrial and Office Properties, 2005 (www.naiop.org/developmentmag /specialsections/200504indexb.cfm).

143. Kats et al., "Green Buildings," 2003.

144. Cohen, personal communication with author, 2007.

145. See www.newresourcebank.com.

146. Greg Kats, the principal author of this book, is a founder of New Resource Bank and helped shape this program.

147. Additional examples from the study data set include the following: (1) a report of more rapid leasing at the Alcyone apartments, in Seattle, Washington; and (2) higher occupancy rates and reduced operating expenses at the Banner Bank Building, in Boise, Idaho.

148. Jeff Martin, Brian Swett, and Doug Wein, "Residential Green Building: Identifying Latent Demand and Key Drivers for Sector Growth" (master's thesis, University of Michigan, Ross School of Business, 2007).

149. John Gattuso, senior vice president, Liberty Property Trust, personal communication with author, 2008.

150. T. W. Chappell, "Case Study: The Louisa," Pacific Security Capital, 2007.

151. The Canyon-Johnson Urban Communities Fund is chaired by former California state treasurer Phil Angelides.

152. The database is used by 75,000 real estate professionals in the United States and the United Kingdom.

153. Spivey, "Commercial Real Estate."

154. Ibid.

155. For a detailed critique and discussion of the CoStar methodology and results, see Scott Muldavin, "Quantifying 'Green' Value: Assessing the Applicability of the CoStar Studies," Green Building Finance Consortium (www.greenbuildingfc.com/Home/Reports.aspx).

156. Ibid.

157. McGraw-Hill Construction, *Green Outlook 2009.*

158. Martin, Swett, and Wein, "Residential Green Building, 2007."

159. Builders of green homes in Terramor Village, in Ladera Ranch, California, reported 5% to 10% price premiums over nongreen homes in Ladera Ranch. (In Martin, Swett, and Wein,

"Residential Green Building.")

160. Martin, Swett, and Wein, "Residential Green Building," 2007.

161. McGraw-Hill Construction, *Green Outlook 2009* (McGraw-Hill Construction, 2008).

162. Ibid.

163. "U.S. and European Investors Tackle Climate Change Risks and Opportunities," Ceres press release, 2008 (www.incr.com/Page.aspx?pid=838).

164. See www.cdproject.net/.

165. Martin Dettling, Albanese Organization, personal communication with author, 2007.

166. Ernst & Young, "Real Estate Market Outlook 2007."

167. Green Real Estate News, December 2007 (see www.greenrealestatenews.com).

168. USGBC, "LEED Initiatives in Governments and Schools," USGBC, updated 2009 (www.usgbc.org/DisplayPage.aspx?CMSPageID=1852).

169. McGraw-Hill Construction, *Global Green Building Trends* (McGraw-Hill Construction, 2008).

170. There is a time lag between project registration and actual construction, meaning that annual LEED registrations do not provide an accurate picture of green construction starts in the year of registration.

171. See Kats, "Greening America's Schools," 2006.

PART II

1. EIA, "Energy Data Book"; "The Building Sector: A Hidden Culprit," Architecture 2030 (www.architecture2030.org/current_situation/building_sector.html).

2. EPA, "Our Built and Natural Environments," EPA, 2001 (www.epa.gov/dced/pdf/built.pdf).

3. Robert Burchell et al., *Sprawl Costs* (Washington, D.C.: Island Press, 2005), 95.

4. Lawrence Frank, Martin Andersen, and Thomas L. Schmid, "Obesity Relationships with Community Design, Physical Activity, and Time Spent in Cars," *American Journal of Preventative Medicine* 27, no. 2 (2004): 87–96.

5. Burchell et al., *Sprawl Costs*, 63.

6. Ibid.

7. Public infrastructure—including roads, sewer systems, public parks, and walkways—can have longer lifetimes than many building systems, so the impacts of green community development are likely to extend for 50 years or more, increasing the value of benefits. However, to maintain consistency with the benefit estimates for green buildings, this book estimates green community benefits for only 20 years—meaning that the estimates are conservative (low).

8. Note that the three dimensions addressed in table 2.2 are analogous to the three categories for which points are awarded in LEED-ND; see www.usgbc.org.

9. For links to resources, organizations, and publications, see EPA's smart growth page: www.epa.gov/dced/.

10. Susan J. Binder, "The Straight Scoop on SAFETEA-LU," *Public Roads* 69, no. 5 (March/April 2006); available at www.tfhrc.gov/pubrds/06mar/01.htm.

11. Shelly Banjo, "You Are How You Live," *Wall Street Journal*, March 24, 2008.

12. See www.cnu.org/node/1440.

13. See www.usgbc.org.

14. Jennifer Henry, USGBC, personal communication with author, 2007.

15. Section 2.2 was adapted and updated from Douglas Farr, *Sustainable Urbanism* (Wiley, 2007).

16. Government Law Center, Albany Law School, "Smart Growth and Sustainable Development: Threads of a National Land Use Policy," spring 2002 (www. governmentlaw.org/files/VLRSmart_growth.pdf), 4.

17. Oregon State Senate, Oregon Land Use Act (SB 100), enacted 1973 (www.oregon.gov/LCD/docs/bills/sb100.pdf).

18. Le Corbusier, *The Athens Charter* (New York: Viking, 1973 [1943]), 54.

19. Ibid., 25.

20. John Norquist, president, Congress for the New Urbanism, speech, January 2004, McLean

County, Illinois.

21. Bill Lennertz and Aarin Lutzenhiser, *The Charrette Handbook: The Essential Guide for Accelerated, Collaborative Community Planning* (Chicago: American Planning Association, 2006).

22. See www.smartcodecentral.com.

23. David Gottfried, *Greed to Green: The Transformation of an Industry and a Life* (Berkeley, Calif.: WorldBuild, 2004).

24. Rob Watson, "What a Long Strange Trip It's Been," PowerPoint presentation, Greenbuild Conference, Atlanta, 2005.

25. Dover, Kohl & Partners, "The Belle Hall Study," 2007 (www.doverkohl.com/files/pdf/Belle%20Hall_low%20res.pdf).

26. Christopher Leinberger, *The Option of Urbanism* (Washington, D.C.: Island Press, 2008).

27. Joseph E. Gyourko and Witold Rybczynski, "Financing New Urbanism Projects: Obstacles and Solutions," Fannie Mae Foundation, 2000 (www.mi.vt.edu/data/files/hpd%2011(3)/hpd%2011(3)_gyourko.pdf).

28. Ibid.

29. Gyourko and Rybczynski, "New Urbanism Projects," 2000.

30. Amita Juneja et al., "Understanding the Concept and Drivers of Mixed-Use Development: A Cross-Organizational Membership Survey" (paper presented at the Conference on Mixed-Use Development, Hollywood, Florida, 2006).

31. Ibid.

32. Craig Q. Tuttle, Jill C. Enz, and Steven I. Apfelbaum, "Cost Savings in Ecologically Designed Conservation Developments," Applied Ecological Services, Inc., 2007.

33. See www.lightimprint.org.

34. National Oceanic and Atmospheric Administration (NOAA), "Alternatives for Coastal Development: One Site, Three Scenarios," NOAA, 2005 (www.csc.noaa.gov/alternatives/).

35. In this analysis, the TND design included slightly fewer units—although all were on smaller lots—yielding 445 acres of open space rather than 85 acres. See www.csc.noaa.gov/alternatives/.

36. Hagler Bailly Services, Inc., and Criterion Planners/Engineers, "The Transportation and Environmental Impacts of Infill versus Greenfield Development: A Comparative Case Study Analysis," EPA, Urban and Economic Development Division, 1999 (www.epa.gov/dced/pdf/infill_greenfield.pdf); Ken Snyder and Lori Bird, "Paying the Costs of Sprawl: Using Fair-Share Costing to Control Sprawl," Smart Communities Network, 1998 (www.smartcommunities.ncat.org/articles/sprawl.shtml).

37. Laurie Volk and Todd Zimmerman, "Development Dynamics: Density and a well-integrated mix of land uses in master-planned communities provide development efficiency and flexibilty," *Wharton Real Estate Review*, vol. II, no. 2 (1998).

38. Leinberger, *Option of Urbanism*.

39. Lynn Richards, "Protecting Water Resources with Higher-Density Development," EPA 231-R-06-001, 2006 (www.epa.gov/dced/water_density.htm).

40. Dover, Kohl & Partners, "Belle Hall Study."

41. Neal Peirce, "An Entire Green Community: Seattle's 21st-Century Model," *Washington Post* Writers Group, 2006. HOPE VI is a federal urban redevelopment project, launched in 1992, that revitalizes deteriorating public housing projects as walkable, mixed-income neighborhoods. See www.hud.gov/offices/pih/programs/ph/hope6/.

42. Chesapeake Bay Foundation, *Our Built and Natural Environments: A Better Way to Grow* (Chesapeake Bay Foundation, 1996), 28–31.

43. Richards, "Protecting Water Resources."

44. Douglas M. Johnston, John B. Braden, and Thomas H. Price, "Downstream Economic Benefits of Conservation Development," *Journal of Water Resources Planning and Management* 35 (January–February 2006).

45. See, for example, L. T. Glickman et al., "Herbicide Exposure and the Risk of Transitional Cell Carcinoma of the Urinary Bladder in Scottish Terriers," *Journal of the American Veterinary Medical Association* (April 15, 2004): 1290–1297; Alberto Ascherio et al., "Pesticide Exposure and Risk for Parkinson's Disease," *Annals of Neurology* 60, no. 2 (2006): 197–203.

46. E. Dumbaugh, "Safe Streets, Livable Streets," *Journal of the American Planning Association* (2005): 71; available at www.naturewithin.info/Roadside/TransSafety_JAPA.pdf.

47. R. Hall, "Walkable Streets: Re-Engineering the Suburban DNA," University of Miami School of Architecture, Program in Community Building, 2001.

48. Dumbaugh, "Safe Streets," 2005.

49. Ibid.

50. Peter Swift, Dan Painter, and Matthew Goldstein, "Residential Street Typology and Injury Accident Frequency" (paper presented at the Congress for the New Urbanism, Denver, Colorado, June, 1997). Additional data added in the summer and fall of 2002.

51. Robert Steuteville and Philip Landon, *New Urbanism: Comprehensive Report & Best Practices Guide* (New Urban Publications, 2003).

52. Federal Highway Administration (FHA), "Status of the Nation's Highways, Bridges, and Transit: 2006 Conditions," FHA, 2006 (www.fhwa.dot.gov/policy/2006cpr/es05h.htm).

53. This estimate uses the same 7% discount rate used elsewhere in the book, and assumes that the cost of injuries rises at a 3% inflation rate—which is lower than the recent historic rise in medical costs.

54. FHA, "2006 Conditions."

55. S. L. Handy, "Understanding the Link between Urban Form and Nonwork Travel Behavior," *Journal of Planning Education and Research* 15 (1996): 183–198; Lawrence Frank, "Land Use and Transportation Interaction: Implications on Public Health and Quality of Life," *Journal of Planning, Education, and Research* 20, no. 1 (2000): 6–22.

56. R. Ewing et al., *Growing Cooler: The Evidence on Urban Development and Climate Change* (Washington, D.C.: ULI–the Urban Land Institute, 2007).

57. There has been considerable debate in the scholarly literature over the extent to which the relationship between land use and travel is causal. Some argue that land use patterns may be merely masking the effect of underlying preferences for neighborhood type and/or travel mode. Between 2005 and 2008, several new studies were released that confirmed the effect of land use on travel behavior—even when preferences are taken into account. See, for example, James F. Sallis et al., "Active Transportation and Physical Activity: Opportunities for Collaboration on Transportation and Public Health Research," *Transportation Research Part A* 38, no. 4 (2004): 249–268.

58. Reid Ewing, Rolph Pendall, and David Chen, *Measuring Sprawl and Its Impact*, vol. 1 (October 2002); available at http://209.85.229.132/search?q=cache:naueVWbGzo0J: www.smartgrowthamerica.org/sprawlindex/MeasuringSprawl.PDF+Measuring+Sprawl+and+Its +Impact,+vol.+1&cd=1&hl=en&ct=clnk&gl=us.

59. J. Holtzclaw et al., "Location Efficiency: Neighborhood and Socioeconomic Characteristics Determine Auto Ownership and Use: Studies in Chicago, Los Angeles, and San Francisco," *Transportation Planning and Technology* 25, no. 1 (2002): 1–27.

60. Reid Ewing and Robert Cervero, "Travel and the Built Environment: A Synthesis," *Transportation Research Record* 1780 (2001): 87–114.; Holtzclaw et al., "Location Efficiency"; Lawrence Frank et al., "Multiple Pathways from Land Use to Health: Walkability Associations with Active Transportation, Body Mass Index, and Air Quality," *Journal of the American Planning Association* 72, no. 1 (2006); L. D. Frank, B. Stone, and W. Bachman, "Linking Land Use with Household Vehicle Emissions in the Central Puget Sound: Methodological Framework and Findings," *Transportation Research Part D* 5, no. 3 (2000): 173–196.

61. Lawrence Frank & Company, Inc.; Mark Bradley; and Keith Lawton Associates, "Travel Behavior, Emissions, and Land Use Correlation Analysis in the Central Puget Sound," report no. WA-RD 625.1, Washington State Department of Transportation, 2005.

62. Ibid.

63. Ibid.

64. L. Frank and J. Chapman, "Integrating Travel Behavior and Urban Form Data to Address Transportation and Air Quality Problems in Atlanta," prepared for the Georgia Department of Transportation (GDOT) and the Georgia Regional Transportation Authority, April 2004.

65. Parsons, Brinckerhoff Quade & Douglas, Inc.; Cambridge Systematics, Inc.; and Calthorpe Associates, *The Pedestrian Environment: Portland, Oregon* (1000 Friends of Oregon, 1993).

66. Ibid.

67. Lawrence Frank & Co. et al., *A Study of Land Use, Transportation, Air Quality, and Health in King County, Washington* (King County Office of Regional Transportation Planning, 2005).

68. L. D. Frank et al., "Urban Form, Travel Time, and Cost Relationships with Work and Non-Work Tour Complexity and Mode Choice," *Transportation* 35, no. 1 (January 2008).

69. S.C. Rajan, "Climate Change Dilemma: Technology, Social Change or Both? An Examination of Long-Term Transport Policy Choices in the United States," *Energy Policy* 34, no. 6 (2006): 664–679.

70. Lawrence Frank, Sarah Kavage, and Bruce Appleyard, "The Urban Form and Climate Change Gamble," *Planning* (August–September 2007).

71. Frank and Chapman, "Travel Behavior," 2004.

72. Reid Ewing et al., "Relationship between Urban Sprawl and Physical Activity, Obesity, and Morbidity," *American Journal of Health Promotion* 18, no. 1 (2003): 47–57; Lawrence Frank et al., "Linking Objective Physical Activity Data with Objective Measures of Urban Form," *American Journal of Preventive Medicine* 28, no. 2S (2005); R. Sturm and D. A. Cohen, "Suburban Sprawl and Physical and Mental Health," *Public Health* 118, no. 7 (2004): 488–496.

73. W. C. King et al., "The Relationship between Convenience of Destinations and Walking Levels in Older Women," *American Journal of Health Promotion* 18 (2003): 74–82; Brian E. Saelens, J. F. Sallis, and L. D. Frank, "Environmental Correlates of Walking and Cycling: Findings from the Transportation, Urban Design, and Planning Literature," *Annals of Behavioral Medicine* 25, no. 2 (2003): 80–91.

74. C. Lee and A. V. Moudon, "Physical Activity and Environment Research in the Health Field: Implications for Urban and Transportation Planning Practice and Research," *Journal of Planning Literature* 19, no. 2 (2004): 147–181; Paul M. Hess, "Pedestrians, Networks, and Neighborhoods: A Study of Walking and Mixed-Use, Medium-Density Development Patterns in the Puget Sound Region" (Ph.D. dissertation, University of Washington, 2001).

75. R. Sturm and D. A. Cohen, "Suburban Sprawl and Physical and Mental Health," *Public Health* 118, no. 7 (2004): 488–496.

76. Frank, Andersen, and Schmid, "Obesity Relationships."

77. Asha Weinstein and P. Schimek, "How Much Do Americans Walk? An Analysis of the 2001 NHTS" (paper presented at the annual meeting of the Transportation Research Board, 2005).

78. L. M. Besser and A. L. Dannenberg, "Walking to Public Transit: Steps to Help Meet Physical Activity Recommendations," *American Journal of Preventive Medicine* 29, no. 4 (2005): 273–280.

79. K. E. Powell, L. M. Martin, and P. P. Chowdhury, "Places to Walk: Convenience and Regular Physical Activity," *American Journal of Public Health* 93, no. 9 (2003): 1519–1521.

80. P. J. Troped et al., "Associations between Self-Reported and Objective Physical Environmental Factors and Use of a Community Rail-Trail," *Preventive Medicine* 32 (2001): 191–200.

81. Estimates of reduced gas use and car ownership, reduced health care expenditures, and increased use of public transport are based on differences in VMT and physical activity in the most versus the least walkable neighborhoods.

82. As this book was being written, gas prices fluctuated widely, from under $2 to over $4 per gallon. Although they were under $3/gallon at the time of publication, they are generally expected to rise over time.

83. See SMARTRAQ gas cost estimates, and Bureau of Labor Statistics, "Table 3023: Selected Southern Metropolitan Statistical Areas: Average Annual Expenditures and Characteristics, Consumer Expenditure" for the Atlanta region (ftp://ftp.bls.gov/pub/special.requests/ce/msa/y0506/south.txt).

84. We can assume that to serve the household member who gives up a car, a maximum of 12 months of public transit passes will be purchased, at a cost of roughly $620 per household per year—versus the Atlanta average for spending on transit, which is $140 per year.

85. Savings estimates for health care costs were prepared in consultation with Melanie Simmons, director of the Healthy Communities Program at Florida State University.

86. The following estimates were based on the results of the Atlanta SMARTRAQ study, and on consultations with Melanie Simmons. See www.cdc.gov/healthyplaces/.

87. Lawrence Frank & Company, Inc.; Mark Bradley; and Keith Lawton Associates, *Travel Behavior, Emissions, & Land Use Correlation Analysis in the Central Puget Sound*, Washington State Department of Transportation, report no. WA-RD 625.1, 2005; Lawrence Frank et al., "Linking Objective Physical Activity Data with Objective Measures of Urban Form," *American Journal of Preventive Medicine* 28, no. 2S (2005).

88. See www.cdc.gov/od/oc/media/pressrel/r2k1006a.htm.

89. L. D. Frank, J. Kerr, and J. Sallis, "Urban Form Relationships with Walk Trip Frequency and Distance among Youth," *American Journal of Health Promotion* 21, no. 14 (2007): supplement, 305. The researchers reported a 99.9% confidence level for their findings.

90. House size was assumed to be the national average (roughly 2200 square feet); see EIA, "Square Footage Measurements and Comparisons" (www.eia.doe.gov/emeu/recs/sqft-measure.html).

91. See www.cnt.org/tcd/ht.

92. Leinberger, *Option of Urbanism*, 2008.

93. Jonathan Levine and Lawrence Frank, "Transportation and Land-Use Preferences and Residents' Neighborhood Choices: The Sufficiency of Compact Development in the Atlanta Region," *Transportation* (2006).

94. Ibid.

95. Jonathan Levine, Aseem Inam, and Gwo-Wei Torng, "A Choice-Based Rationale for Land Use and Transportation Alternatives," *Journal of Planning Education and Research* 24, no. 3 (2005): 317–330.

96. Shyam Kannan, personal communication, 2008.

97. Leinberger, *Option of Urbanism*, 2008.

98. Charles Tu and Mark J. Eppli, "Valuing New Urbanism: The Case of Kentlands," *Real Estate Economics* 27, no. 3 (1999): 425–451.

99. Ibid.

100. Unpublished paper, Applied Ecological Services, 2008.

101. Ibid.

102. National Park Service (NPS), Rivers, Trails, and Conservation Assistance Program, *Economic Impacts of Protecting Rivers, Trails, and Greenway Corridors*, 4th ed. (NPS, 1995).

103. Jeff Lacy, *An Examination of Market Appreciation for Clustered Housing with Permanently Protected Open Space*, Center for Rural Massachusetts Monograph Series (Amherst: University of Massachusetts, August 1990).

104. Ibid.

105. Western Reserve Conservation and Development Council (WRCDC), *Conservation Development and Resource Manual* (WRCDC, 1998).

106. Ibid.

107. Section 2.6 is adapted from Christopher B. Leinberger, *The Option of Urbanism* (Washington, D.C.: Island Press, 2008).

108. The comparative research for these metropolitan areas was gathered from www.realtor.com; see also Christopher B. Leinberger, *The Option of Urbanism* (Washington, D.C.: Island Press, 2008), 97.

109. Leinberger, *Option of Urbanism*, 97.

110. Ibid., 100.

111. Internal RCLCO research undertaken for *Option of Urbanism*; *Option of Urbanism*, 101.

112. "Tysons Corner in 2007 is accessible only by car and rather poor bus service. However, the Metro rail system will serve the area by 2012. Hopes are that the coming of the Metro will convert this sprawling drivable sub-urban place into a walkable urban location, which will be a daunting task given the huge eight-lane streets that bisect it" (Leinberger, *Option of Urbanism*, 101).

113. Jonathan D. Miller, *Emerging Trends in Real Estate 2006* (Washington, D.C.: ULI–the Urban Land Institute, 2006); Leinberger, *Option of Urbanism*, 112.

114. Ibid.

115. Bre Edmonds, "Transit-Oriented Development Sweeps Suburbia," Real Estate Business Online, February 26, 2007 (www.rebusinessonline.com/article_archive/02-26-07.shtml); Leinberger, *Option of Urbanism*, 112.

116. Jane Jacobs, *The Death and Life of Great American Cities*, Vintage House, New York, 1992, p. 7.

117. Peirce, "Entire Green Community"; Linda Baker, "In Tacoma, Recreating Public Housing," *New York Times*, June 24, 2007 (www.nytimes.com/2007/06/24/realestate/24nati.html).

118. See www.cnt.org/ht/.

119. Yan Song and Gerrit-Jan Knaap, "New Urbanism and Housing Values: A Disaggregate Assessment," *Journal of Urban Economics* 54 (2003).

120. Leinberger, *Option of Urbanism*, 2008.

121. Besser and Dannenberg, "Physical Activity Recommendations."

122. J. Kim, "Sense of Community in Neotraditional and Conventional Suburban Developments: A Comparative Case Study of Kentlands and Orchard Village" (Ph.D. dissertation, University of Michigan, School of Architecture, 2001).

123. Podobnik, "Orenco Station."

124. Scott C. Brown et al., "The Relationship of Built Environment to Social Behaviors and Mental Health in Hispanic Elders: The Role of 'Eyes on the Street,'" *American Journal of Public Health* (2007).

125. Arnold R. Spokane et al., "Identifying Streetscape Features Significant to Well-Being," *Architectural Science Review* 50, no. 3 (2007): 234–245.

126. Al Nichols Engineering, "Energy and Water Use in Tucson, April 2006–March 2007: Civano Residences Compared to Tucson's Pre-1996 and 1998/2004 Homes," 2007.

127. Simmons B. Buntin, "Civano: The Dark and the Light: A Debate," Terrain.org: A Journal of the Built and Natural Environments," Fall–Winter 2004 (www.terrain.org/columns/15/literal.htm).

128. See www.appliedeco.com.

129. Randall Arendt, *Conservation Design for Subdivisions* (Washington, D.C.: Island Press, 1996); Steve Apfelbaum and Jack Broughton, "Using Ecological Systems for Alternative Storm-Water Management," *Land and Water* (September–October 1999); Apfelbaum et al., "On Conservation Developments and Their Cumulative Benefits," in *A National Symposium: Assessing the Cumulative Impacts of Watershed Development on Aquatic Ecosystems and Water Quality* (1996); see www.appliedeco.com/Projects/.

130. Steve Apfelbaum, "The Role of Landscapes in Storm-Water Management," Applied Ecological Services, 1993; see www.appliedeco.com/Projects/.

131. In five of the conservation developments, a portion of the additional units are duplexes, townhouses, or condominiums, providing a more diverse mix of products that appeals to different target-market price points and therefore serves a broader demographic.

132. See http://buildingecology.com/free_article_detail.php?id=109&title=Calculating_Buildings %27_Greenhouse_Gas_Emissions_-_%3Cem%3EHal_Levin%3C/em%3E; www.vattenfall.com/www/ccc/ccc/569512nextx/574152abate/574510build/index.jsp.

133. WBCSD, "Energy Efficiency in Buildings," 2007.

134. Ibid.

135. Ibid.

136. See products.bre.co.uk/breeam/index.html.

137. See www.gbcaus.org/greenstar/.

138. See www.hk-beam.org/general/home.php.

139. See www.ibec.or.jp/CASBEE.

140. See www.worldgbc.org.

141. USGBC, LEED project list, January 2009.

142. EIA, "Residential Energy Consumption Survey," 2005 (www.eia.doe.gov/emeu/recs/recs2005 /c&e/summary/pdf/tableus1part1.pdf).

143. See www.passivhaus.org.uk/index.jsp?id=668.

144. Ibid.

PART III

1. U.S. Conference of Catholic Bishops, "Renewing the Earth," 1991 (http://www.usccb.org/sdwp/ejp /bishopsstatement.shtml).

2. World Council of Churches, "Climate Change and the Quest for Sustainable Societies," 1998.

3. John Chryssavgis, ed., *Cosmic Grace, Humble Prayer: The Ecological Vision of the Green Patriarch Bartholomew I* (Grand Rapids, Mich.: Eerdmans, 2003).

4. Issued by the Coalition on the Environment and Jewish Life, Washington, D.C., March 10, 1992.

5. See www.nrpe.org/statements/interfaith_intro01.htm.

6. See www.baptistcreationcare.org/; Jane Lampman, *Christian Science Monitor*, March 12, 2008

(www.csmonitor.com/2008/0312/p02s03-usgn.html).

7. Drew Christiansen, SJ, and Walter Grazer, eds., *And God Saw It Was Good: Catholic Theology and the Environment* (U.S. Conference of Catholic Bishops, 1996).

8. U.S. Conference of Catholic Bishops, "Renewing the Earth: An Invitation to Reflection and Action on Environment in Light of Catholic Social Teaching," statement, U.S. Conference of Catholic Bishops, November 14, 1991.

9. Pope John Paul II, "The Ecological Crisis: A Common Responsibility," January 1, 1990 (www.ncrlc.com/ecological_crisis.html).

10. See www.nrpe.org.

11. Pope John Paul II, "The Ecological Crisis."

12. See www.presentationcenter.org.

13. See www.portsmouthabbey.org.

14. See www.sacredheartmonastery.com.

15. See www.clydemonastery.org.

16. Unless otherwise noted, all quotes were obtained in the course of the research.

17. See www.pittsburghfoodbank.org/.

18. Felician Sisters Convent and Our Lady of the Sacred Heart High School, "Design Cost Data," November–December, 2004.

19. Pauline Dubkin Yearwood, "Judaism Goes Green: Chicago Jewish Organizations Doing Their Part for the Environment," *Chicago Jewish News,* June 22, 2007.

20. See www.fcnl.org/press/releases/silverLEED90407.htm.

21. See http://crs.org/about/guiding-principles.cfm.

22. See www.calvin.edu/academic/geology/.

23. See www.calvin.edu/academic/biology/.

24. See www.calvin.edu/academic/engineering/.

25. See www.calvin.edu/admin/provost/environmental/sustainabilitystatement.html.

26. See www.sidwell.edu.

27. Quinnipiac University poll, 2002; "Patterns of Household Charitable Giving by Income Group, 2005," prepared by Google for Indiana University, Center on Philanthropy, 2006.

28. See "Shades of Green at PLU: Food Sustainability," September 23, 2006 (http://news.plu.edu/node/1168).

PART IV

1. For all model assumptions, please refer to appendix L.

2. For both scenarios, we calculated the compound annual growth rate of energy consumption between 2025 and 2030 and extended those trends out an additional 20 years.

3. EIA, "Annual Energy Outlook 2009," March 2009 (www.eia.doe.gov/oiaf/aeo/). As reported by the EIA, building-sector energy consumption and emissions are essentially the sum of energy consumption and emissions from the commercial and residential sectors. This includes not only the energy used by buildings on site, but also the primary energy used to generate and transport the electricity that the buildings consume. All told, the EIA projects the building sector to represent 41% of total U.S. energy use and 40% of U.S. energy-related emissions in 2010. To these figures, we added (1) conservative estimates for consumption and emissions associated with industrial office space, and (2) the embodied energy and emissions in building materials—which, taken together, increased the estimated building sector share of total U.S. energy use to 45%.

4. Our model assumed that conventional new construction in 2010 would be 15% more efficient than the average existing building; this figure was based on internal discussions at GEI, and on an estimate from Mark Frankel, technical director of the New Buildings Institute. We defined "comprehensive retrofit" as a full upgrade of a building's lighting, HVAC systems, and shell. Our estimate of possible savings from comprehensive conventional retrofits (20% in 2010) is based on internal discussions at GEI, and on a conversation with Steve Nadel, executive director of the ACEEE. We estimated that in 2010, new green construction would be 35% more energy efficient than conventional construction. In the study data set, the average energy savings, when compared with conventional construction, were 36%, and the median was 34%. USGBC data show that LEED-NC 2.2 build-

ings use, on a weighted average, 14% less energy than buildings that meet the ASHRAE 90.1 2004 standards. When buildings that achieved *less* than two points for LEED Energy and Atmosphere Credit 1 are assumed to achieve two points (in keeping with the updated rating system), the average energy savings increase to 17%. We assume that green retrofitted buildings are roughly 25% more efficient than conventionally retrofitted ones.

5. To account for minor upgrades, we assumed that buildings in the existing building stock that are not retrofitted reduce their energy consumption by 0.2% annually. We assumed simple annual straight-line reductions in energy use for each new construction and retrofit type. (Please see appendix N.) These figures are estimates.

6. Our assumption of 5% is based on USGBC data and projections from McGraw-Hill. USGBC figures show that roughly 3 million square feet of LEED homes were certified in 2008 (0.1% of the residential new construction market); 2008 registrations imply roughly 9 million square feet of construction in 2010 (0.3% of residential new construction). McGraw-Hill estimates that in 2010, roughly 15% of new commercial construction will be green. According to the EIA's projections, residential construction is roughly two-thirds of total U.S. new construction. Thus, on a weighted average basis, between 5% and 6% of new construction in 2010 should be green.

7. This is a GEI assumption. LEED for Existing Buildings: Operations and Maintenance is still in too early a stage for the USGBC to have gathered data on market adoption. We can be confident, however, that the figure is very small.

8. LEED-NC and LEED-CS certifications grew, on average, 88% per year between 2003 and 2008. Growth rates slowed sharply at the end of 2008 and in the first quarter of 2009, reflecting a sharp drop in construction across the board.

9. Green retrofits grow at a faster rate than new construction, but from a much lower base.

10. To account for a portion of new construction that, for technical or other reasons, will never be built green, penetration for new green construction never rises above 95%.

11. Penetration for green retrofits never rises above 75%: a significant portion of the market remains content with the still-appreciable savings achieved with standard comprehensive retrofits.

12. Our estimate for the frequency of comprehensive retrofits for the BAU scenario is based, in part, on a conversation with Jennifer Amann, director, Buildings Program, AMEEE.

13. USGBC credit tallies provide some data addressing on-site renewable energy production. In a data set of 338 LEED-NC buildings, the buildings, on a weighted-average basis, produce 1% of the energy that they consume from on-site renewable sources; this translates into roughly 2% of the buildings' electricity consumption, assuming that all the on-site energy generated is for electricity, and that electricity represents between 45% and 50% of the energy consumed by the buildings on site (calculated from EIA, "Annual Energy Outlook 2009"). In a data set of 1,362 LEED homes, the homes produce at least 2.1% of their electricity on site. The wording of LEED credits only allows calculation of minimums. In addition, the methodology by which the LEED homes credit is calculated ensures that the actual percentage is higher. Further, the cost of solar systems is expected to continue to decrease steadily between 2008 and 2010, making it more likely that LEED buildings will deploy them and that they will generate a larger percentage of the electricity they consume.

14. USGBC credit tallies provide the following indicative data: in a data set of 67 LEED-NC buildings, these buildings, on a weighted-average basis, enter into two-year contracts to obtain at least 12% of their electricity from renewable sources; in a data set of 116 LEED for Commercial Interiors spaces, the spaces, on a weighted-average basis, enter into such contracts for at least 22% of their electricity. The wording of the relevant LEED credits only permits the calculation of minimum percentages. Data are not available for green homes or residential projects. To put these figures in perspective, voluntary purchases of renewable electricity accounted for only 0.5% of total retail electricity sales in 2007 (National Renewable Energy Laboratory, "Green Power Marketing in the United States: A Status Report," 11th ed., October 2008 (www.nrel.gov/docs/fy09osti/44094.pdf).

15. To conduct these calculations, we related electricity to primary energy in three steps. First, according to EIA data (EIA, "Annual Energy Outlook 2009,"), electricity represents 74% of the energy consumed by the average U.S. building (including primary energy consumed in the generation and transport of electricity consumed on site). We used this figure in our calculations to relate a building's total primary energy-consumption footprint to the portion related to the electricity it uses (i.e., the total primary energy consumed in the generation and transport of electricity). We assumed that this figure would hold true throughout the forecasted period. However, from the perspective of primary energy consumption, renewable electricity does not replace electricity generated from fossil fuels in a one-to-one ratio: there is significant primary energy lost in the production of the latter,

while the majority of the renewable generation purchased or used by buildings (wind and solar) do not consume fuel to produce power. Thus, in the second step, we accounted for the significant amount of primary energy that is lost in the generation of fossil-fuel-based electricity (but not in renewable generation). In the last step, we accounted for the losses in transmission and distribution using sector-wide data from EIA's "Annual Energy Review 2007," 221 (www.eia.doe.gov/emeu/aer/pdf/aer.pdf).

16. Estimating the impact of a green building's purchase of off-site renewable energy is difficult. First, under LEED, a green building that makes a two-year purchasing contract satisfies the credit for using green electricity, but it is unclear what percentage of these buildings continue their contracts beyond the two years. Second, green electricity from an old renewable asset at a price near or equal to that of conventional generation may be purchased as easily by a conventional building as a green one. By contrast, a green building willing to pay a premium for new green electricity will drive renewable generation that would not exist at a lower price point. For both of these reasons, data on off-site renewables should be viewed as substantially more uncertain than data on on-site renewables. To reflect these uncertainties, we have assumed only one-half the CO_2 impact implied by the renewable energy purchased as green power or as RECs. Again, later in this section, we have assumed only one-half the CO_2 reduction from the green electricity and RECs purchased by green buildings to be attributable to that purchase.

17. In this calculation, we assumed a capacity factor of 20% for on-site renewables (roughly tracking that of U.S.-based solar photovoltaics) and one of 30% for off-site renewables (roughly tracking that of U.S. wind farms).

18. EIA, "Annual Energy Outlook 2009"; DOE, "2007 Buildings Energy Data Book."

19. We conducted these calculations by determining how much primary energy the use of renewable electricity displaces each year—both by avoiding inefficiencies in generation and, in the case of on-site renewables, by avoiding losses in transmission. We then reduced energy consumption for each scenario by this amount before calculating CO_2 emissions.

20. Wind turbines sometimes generate electricity at night, when electricity demand is typically lowest.

21. This corresponds to the percentage of the building sector's energy consumption from renewable sources in 2030, which is roughly 1.5 to 2 times that which is predicted to occur under the EIA scenario (i.e., increasing to 15% to 20% instead of from 7% to 9.5%).

22. We assigned the 20-year NPV of the benefits calculated from our data set to the square feet of green space built or retrofitted each year. We discounted these financial impacts at the same rate (7%), according to the year of construction or retrofit, and then summed them to determine the present value. To determine the NPV of the green construction and retrofits in 2010 in each scenario, we then added the financial impacts to the per-square-foot green premium calculated from our data set.

23. Because the data set is limited to 104 of the 170 total buildings, and because it is skewed to commercial and institutional buildings (since there are relatively few LEED residential buildings), the results overweight commercial buildings and underweight residential buildings, especially detached single-family homes.

24. As noted earlier, the health benefits are not based on our data set but were derived from other studies. Health benefits range from a 20-year present value of $1/sf to $12/sf; we used the average of $6.50/sf. In addition, the societal benefit of lower energy prices (arising from reduced demand) is estimated to be 25% of the savings that accrue directly to the building owner. Societal benefits from CO_2 reductions will vary, depending on the future price of CO_2. We believe the estimate of $1/sf to be reasonable, in a world where CO_2 prices range from $10 to $20 per ton.

APPENDIX A

1. Cathy Turner and Mark Frankel, "Energy Performance of LEED-NC Buildings," 2008 (www.newbuildings.org/downloads/Energy_Performance_of_LEED-NC_Buildings-Final_3-4-08b.pdf).

2. See www.aashe.org.

3. Greg Kats, "Costs and Financial Benefits of Green Buildings," Capital E, 2003; Greg Kats, "Greening America's Schools: Costs and Benefits," Capital E, 2006. "Greening America's Schools" was based on data collected for Greg Kats, Jeff Perlman, and Sachin Jamadagni, "National Review of Green Schools: Costs, Benefits, and Implications for Massachusetts," Capital E, 2005.

4. See www.buildinggreen.com/hpb/index.cfm and www.usgbc.org/LEED/Project/CertifiedProjectList.aspx.

APPENDIX D

1. LEED Registered Projects, April 12, 2007 (www.usgbc.org). The median green premium for K–12 schools in the data set (1.3%) is lower than the median green premium for the data set when K–12 schools (1.8%) are excluded.

APPENDIX F

1. See www.usgbc.org for current statistics on LEED accredited professionals and growth.

APPENDIX H

1. Turner and Frankel, "Energy Performance."

APPENDIX I

1. This appendix is largely based on Turner and Frankel, "Energy Performance"; the full report can be downloaded at www.newbuildings.org. Our study also included a survey conducted among occupants of 12 green buildings, which assessed comfort in terms of indoor air quality, lighting, thermal comfort, acoustics, and overall satisfaction with the building; the results of the survey are discussed in section 1.7, "Health and Productivity Benefits of Green Buildings."
2. The results of the NBI/USGBC study are available at www.newbuildings.org.
3. See Architecture 2030, "The 2030 Challenge" (http://www.architecture2030.org/2030_challenge /index.html).

APPENDIX J

1. Raftelis Financial Consultants, "National Water Rate Survey," 2006. Provided by Peiffer Brandt.
2. Amy Vickers, "Handbook of Water Use and Conservation," Vickers and Associates, 2001; Amy Vickers, personal communication with author, 2007.
3. Benedykt Dziegielewski et al., "Commercial and Institutional End Uses of Water," American Water Works Association, 2000.
4. Peter W. Mayer et al., "National Multiple Family Submetering and Allocation Billing Program Study," Aquacraft, 2004.
5. Kim Shinn, personal communication with author, 2007.
6. District of Columbia Water and Sewer Authority, "Recommended Combined Sewer System Long-Term Control Plan," 2002.
7. Tom Bagely, Boston Water and Sewer Commission, personal communication with author, 2007.
8. New York Department of Environmental Protection, Bureau of Customer Services, "Water/ Sewer Rate History," 2007.
9. Barbara Booth, manager of rates and revenues, Tucson Water, personal communication with author, 2007.
10. Janet Ward, Atlanta Department of Watershed Management, personal communication with author, 2007.

INDEX

Note: Italicized page numbers indicate tables or figures. An "a" following a page number indicates it is contained in the appendix. A "p" following a page number indicates a Perspective article.

A

acoustics, in green buildings, *52*, 56
affordable housing, 131, 200–201a
affordable housing, green, 41–42, 44, *44–45*
affordable-housing industry, 40–41
air pollutant reductions, and walkability index, 114
air quality. *See* indoor air quality; indoor environmental quality (IEQ)
Albanese Organization, 80–81
Aldo Leopold Center, 15, 28–29
American Association of State Highway and Transportation Officials (AASHTO), 107–8
American Council for an Energy Efficient Economy, 67, *68*
American Institute of Architects' Environmental Resource Guide, 101
American Society of Heating, Refrigerating, and Air-Conditioning Engineers, 216–17a, 219a
Apollo Alliance, *68*
assumptions, global, 224–25a
asthma, 42–43
Atlanta, Georgia, 112–13, 117–20, 129
Avanyu, Utah, 8

B

Bank of America Tower at One Bryant Park, 32p
Banner Bank Building, 38
baselines for cost and performance, 5
BAU scenario. *See* Building Green (Green) vs. Business-as-Usual (BAU) scenarios
benefit modeling, 4–6
benefits, direct and indirect, 6
biophilia, 32p
Black & Veatch, 67, *68*
Boston, Massachusetts, 129
Breathe-easy homes, 42–43
Bronx Zoo Lion House, 28
brownfield investments, 79–80p
building construction, Class A, 81
building design, in green communities vs. conventional sprawl, *94*

building green
    benefits of, xvii–xix, 65, 165
    in China, 142p
    as corporate social responsibility, 72p
    cost premiums of early adopters, 11
    costs of, xv, 8–9, 213–14a
    energy use and CO_2 reductions, 14
    estimating employment impacts of, 68–70
    faith-based communities and, 149, 155, 169
    financial benefits, 173
    industry knowledge in, 214a
    moral dimension to, 151
    water and wastewater savings, 33, 36–39
    See also green buildings
Building Green (Green) vs. Business-as-Usual (BAU) scenarios
    benefits, net present value of, *187*
    building green, growth in, *177*
    CO_2 emissions, *172*, 183
    comparison of, xviii–xix
    construction, net present value of, *173*
    energy consumption projections, *180*
    energy efficiency projections, 180
    energy savings, direct, 187
    floor area by type, cumulative, *178–79*
    forecasts for BAU, 175–77
    forecasts for Green, 177–79
    modeling for, 186
    renewable electricity use, *181*
    renewable energy, cumulative installed capacity in, *182*
    renewable energy, growth in demand for in, 182
    retrofits of existing buildings in, 179–80
Building Investment Decision Support (BIDS) tool, 48
building space per person, 141
Business-as-Usual (BAU) scenario. *See* Building Green (Green) vs. Business-as-Usual (BAU) scenarios

C

Calthorpe, Peter, 101
Calvin College, Grand Rapids, 163–64
car accidents, 107–8, 109, 148
carbon caps, 24
carbon dioxide. *See CO_2* entries
Carbon Disclosure Project (CDP), 80

carbon impact of buildings, 26
carbon market, European, 22
carbon neutrality, 26, 32, 72p
car-dominated neighborhoods, 90, 118–20
caring for the planet, 149–50, 160, 163–64
Catholic Campaign on Climate Change, 152
Catholic Children's Health and Environment
    Campaign, 153
Catholic Relief Services (CRS), 162–63, 165–66
CB Richard Ellis (CBRE), 72p
cement production, and CO_2 emissions, 22, 233
    n.41
Center for American Progress, *68*
Center for Building Performance Diagnostics, 48
Center for Neighborhood Technology (CNT),
    121–23
Center for the Built Environment (CBE), 51–52
Center on Wisconsin Strategy, 67
Centers for Disease Control (CDC), 119–20
Cherokee Investment Partners, 79–80p
Chicago, 129
China, 142p, 146
churches, *158–59*
city planning, resident and community needs in,
    131
Civano, Tucson, 133
Class A building construction, 81
clean energy, xvii, xix, 20, 185
climate change, xix, 24
CO_2 emissions
    from affordable housing vs. green afford-
        able housing, 44
    building sector, 141, *184*
    cap-and-trade program, 186
    cap-and-trade programs, 189–90, 233 n.50
    cement manufacture and, 22, 233 n.41
    financial and societal costs of, 174
    generation of, in Chicago Metro Region,
        114
    in Green vs. BAU scenarios, *172*, 183
CO_2 emissions reductions
    green buildings and, xix, 14
    green design strategies and, 89, 113–16
    national strategy for, 22
    Obama's commitment to, xviii
    in renovated buildings, 13
    valuing, 22–23
    voluntary, 80
coastal development alternatives, *105*
Colorado, 99
Comcast building, 75
common good principle, 163
communities, pre-World War II, 90, 94
community design, green
    and affordable housing, 131
    benefits and risks of, for developers, 147
    and CO_2 emissions reduction, 114

commercial sprawl vs., *94*
cost of, 9
costs, illustrative, of conventional develop-
    ment vs., *95*
costs and benefits of, overview, 89–92
defined, 89
design of, xviii, 79p
development approaches to, 96
development options matrix, *93*
health benefits of, 92, 119–20
and transportation emissions, 183
types of, 95p
value of, 189–90
and vehicle-use reduction, 111–13
See also green communities, benefits of
community life, impact of green building on
See also faith-based communities
community-scale projects, implementation of
    green standards in, 133
commuting patterns, 110
See also vehicle-miles traveled (VMT)
Congrès Internationale d'Architecture Moderne,
    99–100
Congress for New Urbanism, 96, 99–100
conservation developments
    benefits of, 140
    and CO_2 reductions, 188
    design layouts for, *136*
    ecological, 95p, 134–35
    first-cost savings, 34, 104–5, 135–37
    open space in, 126
    property values, 125–26
    residence times, 131
    restoration in, 139
    single-development comparison, 137–39
    strategies for, 39, 96
construction, new, 142p, 175
construction and demolition waste, 70–72
construction downturn, xv
construction materials, conventional vs. green,
    173
consumer demand for green homes, 83p
conventional developments
    building design in green communities
        vs, *94*
    car-dominated, 90
    construction materials for, 173
    costs of, 9
    costs of green community design vs., *95*
    design layouts for, *136*
    road design and accident frequency, 107–8
correctional facilities, 200–201a
cost and benefit estimates, 210–12a
CoStar, 76–77, 80
cost comparison for coastal development
    alternatives, *105*
cost differentials of data set buildings, 10

cost-effectiveness of green buildings, 83, 86–88, 168
cost-effectiveness of green technologies, 31–32
"Cost Savings in Ecologically @index2: Designed Conservation Developments," 103

D

Darden, Tom, 79–80p
data-collection sheet, 192a, 193–94a
data-set buildings
    assembly of, 191–92a
    construction and demolition waste diversion, 71
    cost differentials of, 10
    criteria for inclusion, 3–4
    energy savings, advanced, 28
    energy-use reductions, 15, 26–27, 232 n.22
    green premiums for, 9–10, 12
    LEED level of, 211a
    LEED-New Construction buildings compared to, 210–11a
    limitations of, 6–7
    by type, 200–209a
    types, 4
    by year of completion, 5
daylighting, and productivity, 51
Death and Life of Great American Cities, The (Jacobs), 131
Deer Park Buddhist Center, 162
Dell Children's Medical Center of Central Texas, 64
density, in community design, 112, 117
Denver metropolitan area, 128
design layouts for conventional and conservation developments, 136
Detroit region, 128
Dettling, Marty, 80–81
developers, 135, 147
development costs, horizontal vs. vertical, 103
Dorfman, Julie, 161, 168
downtown area redevelopments, 131–32
Duany, Andrés, 101

E

East Little Havana, Miami, 132–33
ecological conservation design, 95p, 139
ecological crisis as moral issue, 151
economic benefits from energy-efficiency measures, 66
economic benefits of green neighborhood design, 43
economic equity, environmental sustainability and, 152–53
economic impact of energy efficiency investments, 70
economic impact of investment in green building improvements, 69

economic recovery, 146p
efficiency-driven reductions, 21
EIA (Energy Information Agency), 175, 180
electrical loads, flattened, green buildings and, 183–84
electrical supply, conventional, decarbonization of, 186
electricity, average price of, 19
electricity use, and CO_2 emissions, 141
electrochromic glazing, 30
emissions, impact of policy measures on, 185
emissions reductions, xviii, 24–25, 171
    See also CO_2 emissions reductions
employment impacts of building green, xvii, 66–72, 236 n.124
energy, clean, xvii, xix, 20
    See also renewable energy
energy consumption, primary, 245 n.15
energy consumption by buildings, 141, 180
energy efficiency
    cost of, 21, 215a
    economic benefits from, 66
    employment impacts of, xvii, 66–67, 70, 236 n.124
    and energy-use reductions, 32p
    green buildings compared to, xix
    in health care sector, 63
    and property value of green buildings, 81
    standards for new construction in China, 142p
    voluntary vs. mandatory approach, 141
Energy Information Agency (EIA), 175, 180
energy performance, 121p, 218–19a
Energy Policy Act (1992), 34
energy prices, 19–20, 232 n.32
energy savings
    for acute care hospitals, 63
    advanced, for data-set buildings, 28
    in buildings with advanced energy-use reductions, 31
    by building type, 17–19
    direct, 15–20, 53p
    embodied, 22, 71
    for green affordable housing, 42, 44, 45
    in green buildings, 14, 26
    in green schools and offices, 84–86
    indirect, 20–21
    OHSU Center for Health and Healing, 64
    potential, for electrochromic glass used with daylighting controls, 30
    ranges of, per square foot, 17
    water-related savings and, 38
Energy Star buildings, 76, 77
energy use, 53p, 142p, 176, 216–17a, 219a
energy-use reductions
    and CO_2 reductions, 14

for data-set buildings, *15, 26–27, 210*a, 232
n.22
energy efficiency and, 32p
in health care facilities, 64
in LEED for New Construction
buildings, *210*a
by LEED level, *16*
secondary impact of, 20–21
energy users, industrial, 22
Environment Technical Advisory Group, 185
environmental policy in U.S., 98
Environmental Resource Guide (AIA), 101
environmental responsibility, 153
environmental sustainability, and social and economic equity, 152–53
Envision Utah process, 101
equity value, in green affordable housing, 44
ethical and religious values, 190
European carbon market, 22
European Union, and zero net energy homes,
144
Evaluation Standard for Green Building, China,
142p

F

faith-based communities
benefits and potential of, 165
and building green, 149, 155, *156–59*, 169
interview instrument for, 154
responses to environmental conditions, 152
and stewardship of the planet, 160, 163
Faubourg Boisbriand, 79p
Felician Sisters, 160–61, 165, 167
financial benefits of green design, 173, 188
financial costs of CO_2, 174
financial crisis, as opportunity, 146p
financial impacts of greening, 80
financial institutions, and affordable-housing industry, 40–41
financial payback of building green, xvii
financial stewardship, and green initiatives, 168
financial value, estimating, of storm-water reduction, 107
first-cost green premiums in health care, 59, 60–
61, 62–63
first costs, 11, 103–5, 135–37
first-cost savings, xvi, 34, 41, 104–5, 135–37
Fisk, William, 57–58
Forest Stewardship Council-certified wood, 54p
form-based zoning, 101
Franciscans, 160
Francis of Assisi, Saint, 160
Friends Committee on National Legislation
(FCNL), 162, 166–67
Furman University, 166
future development, the shape of, 129–30
future energy prices, 232 n.32

G

gas-fired turbines, 32p
gasoline costs in least and most walkable
neighborhoods, *119*
gas use and car ownership in walkable
neighborhoods, 118–19
gentrification, 131–32
geothermal heat pumps, 29
GHG (greenhouse gas) emissions, xviii, 72p,
113, 171
Glendening, Parris, 99
global assumptions, 224–25a
global financial crisis as opportunity, 146p
Global Green Building Trends, 8
global warming, xix, 24
green affordable developments, costs and
benefits of, 41
green affordable housing. *See* affordable housing,
green
"Green Book" (AASHTO), 108
Green Branch banks, 53–54p
green building data set. *See* data-set buildings
green buildings
acoustics in, *52,* 56
and CO_2 emissions reductions, 14
common features of, xvii
construction of, and job creation, 69
cost-effectiveness of, 11–13, 83, 168
costs, perceived vs. actual, xv
in data set, 3–4
and electrical loads, flattened, 183–84
energy efficiency compared to, xix
energy savings in, 26
health benefits of, 56–58, 119–20
international, *144–45*
lighting in, *52*
productivity benefits of, 51–56, 56–58
and property values, 74–76, 78–79, *81*
sales of, 77–78
simple payback for, 86–88
standards for, 133, 142p, 143 (*See also*
LEED *entries*)
surveys on costs of, 8–9
surveys on value of, 73
value of, 189–90
See also building green
green building survey instrument, 223a
green communities, benefits of
community-wide, 106
for developers and municipalities,
147–48
health improvements, 43
transportation, value of, 118–19
when compared to conventional
sprawl, *94*
Green Communities Criteria (GCC), 40
green design

as branding strategy, 145
goals of, 94
impacts of, 165, 167
rapid transition to, 189–90
risks of not investing in, 80p
green factors, as priorities in home choice, 83p
green goals, early integration of, and cost-
effectiveness, 10
Green Guide for Health Care, 61
greenhouse gas (GHG) emissions, xviii, 72p,
113, 171
green housing, demand for, 78
greenness, dimensions of, 93
Green New Deal, 146p
Green Outlook 2009 (McGraw-Hill), 78
green power. *See* clean energy
green premiums
for data-set buildings, *10, 12, 28*
defined, 9–10
for faith-based projects, 155
for green affordable housing, *45*
for green renovations, 13
limitations of, as concept, 213–14a
green renovations, 13, 167–68
green roofs, 38–39
Green Sanctuary status, 165
Green scenario. *See* Building Green (Green) vs.
Business-as-Usual (BAU) scenarios
ground source heat pumps (GSHPs), 29

H
health benefits
access to public transit and, 132
estimating the value of, in green buildings,
56–58
of green communities, 43, 92, 117–20
IEQ and, 32, 46–51
of walkable neighborhoods, 43, 119–20
health care, 65
health care costs, 119–20, *148*
health care facilities, *59*, 60–61, 64, 200–201a
health care projects, green, defined, 61–62
health care sector, 59, 61–65
heating and cooling systems, conventional, 29
heat pump technologies, 29
heavy metals, 24–25
Heifer International Headquarters, 38
higher education buildings, 200–201a
High Point HOPE VI development, Seattle,
42–43, 100, 107
homebuyer preferences, 78, 124–25
homeowner attitudes toward green residential
products, 83p
home weatherization, 44
HOPE VI program, 42–43, 100, 107
hospitals, 35, 51, 61, 63
houses of worship, *158–59*

housing, 43, 78, 100, 125, 200–201a
See also affordable housing; affordable housing,
green

I
IEQ. *See* indoor environmental quality
incentive programs, market-based, in China,
142p
incentives for PV installations, 30
incentives to support green features, 231 n.14
inclusionary zoning, 132
India, 146
indoor air quality, *49, 52,* 57
See also indoor environmental quality (IEQ)
Indoor Air Quality Scientific Findings Resource
Bank, 21, 48
indoor environmental quality (IEQ)
in China, 142p
at Felician Sisters convent and school, 165
in green affordable housing, 42
health and productivity improvements, 32,
46–51
and workplace satisfaction, 75
industrial efficiency investments, 21
industrial energy users, 22
industry surveys and analysis, 76
infill development, 106
infrastructure, 36, 104–6, *148*, 238 n.7
input-output model, 68
integrated design, 102
Intergovernmental Panel on Climate Change, 24
international green buildings, *144–45*
irrigation water use, 35
Islamic values, 167

J
Jackson, Henry (Scoop), 98
Jacobs, Jane, 131
Jampa, Ani, 162
Jesuits, 166
Jewish Reconstructionist Congregation, 161, 168
jobs. *See* employment impacts
John Paul II (pope), 151
Judaism, 151, 161, 168
Judson University, 166

K
K-12 schools, 202–5a
Kentlands, Maryland, 132
King County, Washington, *115–16*

L
laboratories, 204–5a
Land Use, Transportation, Air Quality, and
Health (LUTAQH) study, Seattle area, 112, 113
Land Use, Transportation, and Air Quality
study, Portland, 113

land use mix, 112, *115*, 117
land use policies, and GHG emissions, 113
land use-travel relationship, 240 n.57
Laurel Springs, Bainbridge, Ohio, 126
Lawrence Berkley National Laboratory, 21, 48
LEED (Leadership in Energy and Environmental
    Design), 47, 102, 143, 231 n.3, 231 n.6
LEED Accredited Professionals, 214a
LEED-certified buildings, 5, 16, 56, 73, 76, 77
LEED for Commercial Interiors, 79p
LEED for Existing Buildings, 13
LEED for Homes, Platinum projects, 79p
LEED for Neighborhood Development, pilot
    program, 79p, 82, 96–97, 102
LEED for New Construction, 73–74, 96, *211a*
LEED Platinum buildings, 75–76, 79p,
    121–23p
LEED registrations, 61–62, 82, *82*, 238 n.170
Lewis and Clark State Office Building, 38
Liberty Property Trust, 75
life-cycle benefits, 4
lighting, in green buildings, *52*
logging practices, unsustainable, 142p
Louisa, The, Portland, Oregon, 76
low-density suburban development, 90–91
low-e windows, 29–30
low-flow toilets, 38
low-income communities, structural disadvan-
    tages of, 133
low-income housing, 43
See also affordable housing; affordable housing,
    green
low-income populations, and walkable
    communities, 132

M

market-transformation strategies, 40
Maryland, 99
Mazzarino, John, 79–80p
McGraw-Hill, 73, 77–78
medical building types, 60
methodology, 3–7
mixed-use developments, 95p, 100–101, 104, 126,
    131–32
modeled energy use, 219a
models
net-present-value (NPV), 4
moral dimension to building green, 151, 160
Morken Center for Learning and Technology,
    166
multifamily affordable housing, 200–201a
multifamily housing, 125
multifamily/mixed-use buildings, 208–9a
municipalities, costs and benefits of green
    community design for, 147
municipal regulations. *See* zoning codes
Muslim Khatri Association, 167

N

National Council of Churches of Christ, 153
National Environmental Protection Act (NEPA),
    98
National Household Transportation Survey, 132
National Land Use Policy Act, 98
National Mainstream GreenHome, 79–80p
National Religious Partnership for the Environ-
    ment, 152–53
natural gas, *19*, 21
nature preserve, Calvin College, 163
neighborhood design, *94*, *111*, 131
neighborhoods, green, 43, 110
neighborhood walkability, 117–18
See also walkability index
New Buildings Institute, 56
New Jersey, 99
new urbanism, 96, 100–101, 127
New York City, 38, 128
nitrogen oxides (NOx), 24–25
NOx (nitrogen oxide) emissions, 24–25

O

Obama, Barack, xviii, 171
obesity rates, 117–18
occupant comfort and satisfaction, 51–52,
    53–54p, 55, 63–64, 121p
office buildings, 38, 73–74, 77, 84–86, 129,
    204–7a, 236 n.115
off-site energy sources, 181–82, 246 n.16
OHSU. *See* Oregon Health Sciences University
    (OHSU)
One Crescent Drive, 75
on-site renewable energy, xviii, 32p, 66–68, 181,
    245 n.13
on-site wastewater treatment, 35p
open space, 118, 126
Orchard Village, Maryland, 132
Oregon Health Sciences University (OHSU)
    Center for Health and Healing, 35p, 64
Orenco Station, Portland, 132
Our Lady of the Sacred Heart School, 160–61
outdoor views, and patient recovery time, 51
outpatient facilities, 61

P

Pacific Lutheran University, 166
particulate matter, 24–25
Passivhaus principles, 143
pedestrian safety, 107–9
perceived risk of TNDs, 104
performance space, 206–7a
Phoenix, 129
photovoltaic installations, 29–30
physical activity, impact of green communities
    on, 117–18
Piraino, Dave, 163, 166

planning and zoning regulations. *See* zoning codes

Playa Viva development, Mexico, 145

PNC Bank, 53–54p, 55

Portland, Oregon, 35p, 76, 113, 132

post-occupancy evaluations, 55–56, 64, 218–19a

productivity benefits in green buildings, 32, 46–51, 51–58

property values
    compared, 125–26
    energy efficiency and, *81*
    for green buildings, 74–76, 78–79
    storm-water flow and, 107
    in walkable urban places, 129

public buildings, other, 206–7a

public housing revitalization, 100

public sector costs, TNDs and, 106

public transit, health benefits of access to, 132

PV installations, 29–30

R

"Real Estate Market Outlook 2007" (Ernst & Young), 73, 81

ReBuilder's Source, 71–72

REBusiness Online, 130

recession, global, xv

recreational facilities, 118

recycled materials, in LEED, 22

recycling of C&D waste, 70–71

Redevco, 145–46

reform movements, 98

Religious Society of Friends (Quakers), 162

Remlick Hall Farm, 107

renewable electricity use
in Green vs. BAU scenarios, *181*

renewable energy
    cost of energy-efficiency and, 215a
    cumulative installed capacity in BAU vs. Green scenario, *182*
    development of new, 185–86
    growth in demand in Green scenario, *182*
    off-site generation, 246 n.16
    on-site generation, xviii, 32p, 66–68, 181, 245 n.13

renewable-energy certificates, 181

Renewable Energy Credits, 182, 185–86

renewable-energy measures, 215a

renewable-energy programs, *68*

research and case studies, 132–33

residential buildings, 208–9a

residential density
and storm-water runoff, 106
and transport-related CO_2 per person, *116*

residential energy efficiency, funding for, 190

restoration, in conservation developments, 139

retail buildings, 208–9a

retrofits of existing buildings, 175, 179–80, 190, 245 n.9, 245 n.11

reuse vs. demolition of existing buildings, 13

Rio Earth Summit, 101, 102

risk, green development and, 80–83, 80p, 104, 147

Robert Charles Lesser & Company (RCLCO), 83p, 124–25

S

Sage electrochromic glass, 29

Saxenian, Michael, 158

Saulson, Gary Jay, 53–54p

schools and academic buildings, faith-based, *156–59*

schools and academic buildings, green, 11, 51, 84–86, *85*, 202–5a

Seattle, Washington, 42–43, 100, 107, 112, 113

sick-building syndrome, 57

Sidwell Friends School, 158, 164

single-family detached homes, 125

site design, and VMT, 113

Smart Code, 101

smart growth, 95p, 96, 98–99

Smart Growth and Neighborhood Conservation Program, 99

SMARTRAQ (Strategies for Metropolitan Atlanta's Regional Transportation and Air Quality), 112–13, 117–18, 120

SO_2 (sulfur dioxide) emissions, 24–25

social cohesion, green community design and, 132–33

social equity, environmental sustainability and, 152–53

societal costs of CO_2, 24, 174

solar photovoltaics (PV), 29–30

source list, 195–98a

sprawl, 90–91, *94*, 111, 117

statewide land use planning, 98–99

stewardship, financial, and green initiatives, 168

stewardship of the planet, 149–50, 160, 163–64

storm-water flow, 106–7

storm-water infrastructure, 39

storm-water management, in Token Creek designs, 138

storm-water reduction, estimating value of, 107

storm-water reduction strategies, 33, 38–39

storm-water treatment train (STT), 138

street connectivity, and transport-related CO_2, *116*

study data set. *See* data-set buildings

study participants, solicitation of, 191a

suburban neighborhoods, standard, 124–25

sulfur dioxide (SO_2), 24–25

survey instrument, green building, 223a

survey instrument for faith groups, 154

sustainable urbanism, 95p, 97

T

technologies, green, 28–30, 31–32
TNDs. *See* traditional neighborhood developments (TNDs)
Token Creek Conservancy Estates, 136–39
town planning, traditional, 100
traditional neighborhood developments (TNDs)
    consumer preference for, 124–25
    development costs, 103
    infrastructure savings, 104–5
    and new urbanism, 96
    perception of higher costs for, 104
    property values in, 125
    public sector costs, 106
    residence times in, 131
    and storm-water flow, reduced, 106–7
    streets and roadways in, 108
    and sustainable urbanism, 95p
transit-oriented development (TOD), 95p
transit-oriented villages, 100–101
transit service, 113, 117
transportation, and green community design, 92, 118–19, 131, *148*, 183
See also SMARTRAQ
transportation energy intensity, CNT and, 122–23p
transport-related CO_2 per person, King County, Washington, *115–16*

U

Unitarian Universalist Church, 165
University of California, Berkeley, 51–52
University of Montreal, 55–56
University of Scranton, 166
urban growth boundaries, 99
Urban Land Institute (ULI), 100, 130
urban-rural transect, 101
U.S., potential impacts of green design in, 102, 171–74
U.S. Conference of Catholic Bishops (USCCB), 152, 162–63
U.S. Green Building Council (USGBC), 16, 78, 101–2

V

vehicle-miles traveled (VMT)
    community design and, 112–13
    fatalities per 100,000 million, 108–9
    green community design and, 110
    home-based, in Atlanta Region, *112*
    sprawl and, 111
    transit service and, 113
    walkable neighborhoods and, 112
venture-capital investment in clean energy, xvii

Vincent and Helen Bunker Interpretive Center, 163–64
vintage building energy consumption, 175
VMT. *See* vehicle-miles traveled (VMT)

W

walkability index, and air pollutant reductions, 114
walkable communities, 96, 132
walkable mixed-use developments, 131–32
walkable neighborhoods
    consumer preference for, 124–25
    gas use and car ownership in, 118–19
    health benefits of, 43
    health benefits of green buildings in, 119–20
    health care costs in, *120*
    physical activity and obesity in, 117–18
    and VMT, 112
    zoning codes as barriers to, 130
walkable urbanism, 95p, 127–30
Washington, D. C., metro area, 128–29
wastewater savings, benefits of, 36–39
wastewater treatment rates, Portland, 35p
water conservation programs, 37–38
water infrastructure, 36
water rate increases, 36, 220–21a
water-related savings
    assumptions for calculations of, 220–21a
    benefits of, 36–39
    by building type, *34*
    cost of, in data-set buildings, 222a
    costs and benefits of, for green schools and offices, 84–86
    and energy savings, 38
    in health care sector, 64
    indirect, 37–38
    by LEED level, 34
    OHSU Center for Health and Healing, 35p
    strategies for, 33–35
    value of, by building type, *36*
water-saving features in data-set buildings, 222a
water sources, new, cost of, 37–38
water use in buildings, 53p, 142p, *221a*
weatherization of homes, 44
Western Reserve Conservation & Development Council, 126
Work Environments Research Group, 55–56
World Business Council for Sustainable Development, 8–9
World Council of Churches, 151
World Green Building Council, 143
World Vision, 152

Z

zero-energy buildings, xviii, 32
zoning codes, 39, 100–101, 104, 129–30, 132, 190